proceedings of the workshop

Intrinsic Multiscale Structure and Dynamics in Complex Electronic Oxides

the
abdus salam
international centre for theoretical physics

united nations educational, scientific
and cultural organization

international atomic
energy agency

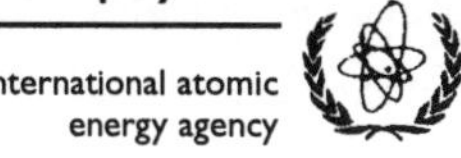

Editors

A. R. Bishop
Los Alamos National Laboratory, USA

S. R. Shenoy
ICTP, Italy

S. Sridhar
Northeastern University, USA

proceedings of the workshop

Intrinsic Multiscale Structure and Dynamics in Complex Electronic Oxides

The Abdus Salam International Centre for Theoretical Physics   1–4 July 2002

World Scientific
New Jersey • London • Singapore • Hong Kong

Published by

World Scientific Publishing Co. Pte. Ltd.

5 Toh Tuck Link, Singapore 596224

USA office: Suite 202, 1060 Main Street, River Edge, NJ 07661

UK office: 57 Shelton Street, Covent Garden, London WC2H 9HE

British Library Cataloguing-in-Publication Data
A catalogue record for this book is available from the British Library.

The image on the front cover reflects the local electronic structure of a 15 nm square field of view of the surface of the high-T_c superconductor, BiSrCaCuO. The colours represent the magnitude of a gap in the density of electronic states which is associated with superconductivity. Black/blue/green/orange/red represent energy gaps in the ranges 70/60/50/40/30 millielectron volts. Clearly the electronic structure of the material varies dramatically over distances less than 1 nm — the key result.

Cover graphic by S. Uchida (Tokyo University) and J. C. Davis (Cornell University); adapted from Nature, Vol. 415, p. 412 (2002).

INTRINSIC MULTISCALE STRUCTURE AND DYNAMICS IN COMPLEX ELECTRONIC OXIDES

ISBN 981-238-268-2

This book is printed on acid-free paper.

Printed in Singapore by Mainland Press

PREFACE

Complex electronic oxides have, in recent years, been found to exhibit patternings in charge, spin and lattice spacings. These patterns involve nanoscale phase separations, and have been variously referred to as stripes, conductive channels, or droplets. Signatures of these surprising spatial variations show up in a variety of high-resolution spatial and temporal probes, including diffuse X-ray and neutron scattering; acoustic and electromagnetic response; NMR and NQR; ultrafast pump-probe spectroscopy; and scanning tunneling microscopy. The ubiquity of such intrinsic inhomogeneities, that range in scale from a few to hundreds of lattice spacings, as well as the importance of the materials that exhibit them, such as superconducting cuprates and colossal magnetoresistance manganites, has motivated intensive experimental and theoretical research into details of the phenomena, the puzzle of their origins, and their possible functional consequences.

This volume contains the Proceedings of a Workshop on "Intrinsic Multiscale Structure and Dynamics in Complex Electronic Oxides", held at the Abdus Salam International Centre for Theoretical Physics, Trieste, Italy, during July 1–4, 2002. The contributions are roughly in the order of the programme. An author index is provided at the back.

We are grateful to all the invited speakers who presented their research at the Workshop, and who contributed to this Proceedings volume. We especially thank Prof. K. A. Müller for his insightful opening presentation, and all participants and chairs who made the three informal discussion sessions such an instructive exchange of ideas. Support for the workshop from ICTP, from Los Alamos National Lab, and from the US Office of Naval Research, is gratefully acknowledged. We thank the Workshop Secretary Ms. Rosanna Sain, for her invaluable help.

We hope this volume of Proceedings will serve as a stimulating reference for those working in this active and fascinating area of Condensed Matter Physics.

A. R. Bishop (Los Alamos National Lab)
S. R. Shenoy (ICTP)
S. Sridhar (Northeastern University)

CONTENTS

ESSENTIAL HETEROGENEITIES IN HOLE-DOPED CUPRATE SUPERCONDUCTORS

K.A. MÜLLER
Physics Institute, University of Zurich, CH-8057 Zurich, Switzerland

A review of recent experiments in hole-doped cuprates is presented. These include photoemission, inelastic neutron scattering, EXAFS, PDF's, electron paramagnetic resonance and susceptibility data. For doping concentrations below optimum, all are compatible with Jahn-Teller bipolaron formation at the pseudogap temperature T^*, with a simultaneous presence of fermionic quasi-particles. The theoretically derived superconductivity onset and maximum T_c at optimal doping agree qualitatively with observation. Very large isotope effects at T^* and lower doping supports the vibronic character of the ground state.

1. Local distortions

The essential structure of high T_c superconductors is well-represented by Tokura's [1] picture of donor/acceptor/inert layers bracketing a CuO_2 sheet conductor. The manybody wave function in the material is vibronic, i.e. includes both nuclear and electronic parts,

$$\Psi = \psi_n \psi_e, \tag{1}$$

without being in the adiabatic slaving limit. The nuclear part ψ_n is probed in neutron scattering experiments, while the electronic part ψ_e is probed by photoemission. There are in general, two types of excitation energies, that are doping-dependent. This "vibronic" picture, in the sense of an interplay between the electronic charge and the ionic lattice structure, is a major theme of this talk.

Photoemission (ARPES) data by Lanzara et. al. [2] clearly shows a common feature in different high T_c superconductors, that is a signature of the lattice playing a role. The quasiparticle energy versus (rescaled) wavevector plots for Bi2212, Bi2201 and LSCO show a kink, while NCCO does not show any such behaviour. The kink occurs near 70 meV at a characteristic wavevector half in the Brillouin Zone, and indicates two different group velocities. They are due to two different quasi-particles.

Probing the other part of the wavfunction in Equ (1), inelastic neutron scattering by Egami and collaborators [3] shows that LO phonon spectra in $YBCO$ and $LSCO$ change significantly with oxygen concentration x varying from 0.95 to 0.20. There is a distinct feature in the dispersion at $60-80$ meV that also occurs in the middle of the Brillouin zone. The intensity of the excitation above the anomaly changes significantly in favour of the one below it. This reflects a change of the ratio of the two types of quasi-particles present.

While structure clearly is expected to play a role, there was already early evidence by $EXAFS$ measurements of Bianconi et. al. [4] that this lattice structure could be *locally varying.* They suggested in $x = 0.15$ $LSCO$ that nano-domains ocurred, with alternating bands or "stripes", separating charge-rich and charge-poor regions. The stripes consist of distorted unit-cell bands (D) of width $\sim$ 8 Angstroms and undistorted unit-cell bands (U) of width $\sim$ 16 Angstroms. The U regions are locally LTO-like, while the D regions have LTT-like CuO_6 octahedra, relatively tilted by about 16 degrees, and with bonds relatively closer in length. We note that the distorted octahedra are similar to a 'Q_2' local mode, familiar in the Jahn-Teller effect, a point we will return to, later.

Charge inhomogeneities are associated with the local patterns of octahedral tilts, or more generally, with a distribution of such tilts as found by the the Billinge group [5] (PDF's). They find from neutron diffuse scattering, that for $LSCO$, the $x = 0.1$ data corresponding to '3 degree tilts' can in fact, be reproduced by combinations of heavily tilted (5 degree) and untilted (0 degree) octahedra.

EPR data by Kochelaev et. al. [6] show that in $LSCO$ that distortions exist. They have been successfully modelled as a 3-spin polaron, again associated with a 'Q_2'-like Jahn-Teller distortion.

In short, a variety of probes, including neutron scattering, PDF, ESR and $EXAFS$ show that structural modes are important at various *finite wavevector* values, involving combinations of JT distortions.

2. Pairing

The intersite JT correlations can in fact, lead to a finite wavevector $q_c = 2\pi/d$ pairing interaction, with a small pair size and short coherence length, $d \sim \xi$.

Kabanov and Mihailovic [7] have considered a model where these ideas lead to an interaction with a coupling constant of the form

$$g(q) = \frac{g_0}{[(q - q_c)^2 + \Gamma^2]}, \tag{2}$$

that is resonant in the wavevector. Furthermore, one can have coupling between $q \neq 0$ phonons and s, d two-fold degenerate electron states as well as coupling to spins, all with the resonant coupling structure of Equ (2).

We note here that Weisskopf had shown for classical superconductors, that

$$\Delta \sim E_F \lambda / \xi, \tag{3}$$

where λ is the screening length. This implies that the superconducting gap/temperature $\Delta \sim T_c$ is high when the coherence length ξ is small. Thus a small-pair model can be a high-T_c model.

In fact, if pairs are small, then superconductivity can be established through some kind of percolation, with pair size ℓ_p smaller than the coherence length, and bigger than the lattice scale a:

$$a < \ell_p < \xi. \tag{4}$$

The picture that emerges is of Jahn-Teller induced *mesoscopic pairs*, that *fluctuate and percolate* [8]. A further development of these ideas yields
i) an understanding of the minimum coherence length observed;
ii) the right percentage of holes for the onset of cuprate superconductivity ($\sim$ 6 percent);
iii) the right ${T_c}^{max}$ value.

3. Pseudogap

Returning to structure, we consider the pseudogap temperature T^* as where the local distortions begin to occur. Work by Bussmann-Holder et. al. [9] has shown that bands of different symmetries $x^2 - y^2, 3z^2 - r^2$ can be coupled by tilting octahedra. The model yields both a T_c and a pseudogap temperature T^*. The pseudogap formation drastically reduces the energy separation between charge and spin levels, from ~ 2 eV to $\sim$ meV.

There are also other related analyses [10] based on the JT polarons combining to form bipolarons, whose binding energy is found to fall as $\sim 1/x$ with doping. So the temperature-doping phase diagram would show a bipolaron formation temperature $T^*(x)$, that decreases with doping x; and is above another negative-slope characteristic temperature, where bipolarons cluster to form stripes. The decreasing T^* line would nearly meet the increasing superconducting T_c phase boundary at the optimum doping, i.e. maximum temperature T_c^{max} for superconductivity.

The local structure, and its formation temperature T^* can be probed by $XANES$ methods, when an X-ray photon ejects an electron from Cu^{2+} and the electron waves interact with the O^{2-} neighbours. Plots in an early paper by Lanzara et. al. [11] show fluorescence counts versus phonon energy, with features that are linked to the neighbouring La/Sr, and in-plane oxygen. NQR is another important probe of local structure, as used by Imai and collaborators [12].

The temperature-dependence of the $XANES$ peak intensity ratio shows a dip at a $T^* \sim 160$ K, that is associated with stripe formation. Since this technique probes only oxygen neighbours, it is a site-specific way to investigate effects of isotopic substitutions. In fact there is a large isotope effect, with $O^{(16)} \rightarrow O^{(18)}$ substitution causing a rise to $T^* \sim 180$ K.

Several techniques can be used to probe the T^* versus doping curve for $La_{2-x}Sr_xCuO_4$, and the techniques like XANES,NQR, and EPR probe widely different time scales ($\sim 10^{-13}sec, 10^{-7}sec, 10^{-9}sec$). Nonetheless, the data fall on the same curve.

Other investigations include inelastic time-of-flight neutron scattering measurements by Rubio Temprano et. al. [13] that substitute for both copper and oxygen. This work, on the isotopic series $HoBa_2Cu^{(n)}{}_4O^{(p)}{}_8$ where $n = 63, 65$ and $p = 16, 18$, reveals a large T^* isotope effect for oxygen ($\alpha^*{}_O = -2.2$), and an even larger value for copper ($\alpha^*{}_{Cu} = -4.9$). Thus *both* copper and oxygen play a role in stripe formation, on a scale $\sim 10^{-14}sec$, that cannot be accounted for by magnetic interaction.

4. Concluding comments

Most theories of cuprate superconductors use the machinery of highly correlated conducting matter, and in them the lattice is taken as rigid at the outset. The many experiments of quite different character outlined here tell a quite different story, especially if the time scale of a particular experiment is sufficiently short. Then the heterogeneity becomes manifest in a clear manner. For all of them, the vibronic character of the ground state is present in a consistent way. This however was the concept which led to the discovery of the HTS cuprates [14]. Finally, it should be noted that the JT bipolaron quasi-particles include high correlations as well, which is often overlooked.

Acknowledgements: The author is indebted to S.R. Shenoy, A. Saxena and T. Lookman for compiling and editing the manuscript, based on his oral presentation in Trieste.

References

1. Y. Tokura and T. Arima, Japan J. Appl. Phys. **29**, 2388 (1999)
2. A. Lanzara, P.V. Bogdanov, X.J. Zhou, S.A. Kellar, D.L. Feng, E.D. Liu, T. Yoshida, H. Eisaki, A. Fujimori, K. Kishio, J.I. Shimoyama, T. Noda, S. Uchida, Z. Hussain and Z-X. Shen, Nature **412** 510 (2001)
3. R.J. McQueeney, Y. Petrov, T. Egami, M. Yethiraj, G. Shirane and Y. Endoh, Phys. Rev. Lett., **82**, 628 (1999); J-H. Chung, T. Egami, R.J. McQueeney, M. Yethiraj, M. Arai, T. Yokoo, Y. Petrov, H.A. Mook, Y. Endoh, S. Tajima, C. Frost, and F. Drogan, Phys. Rev. B **67**, 014517 (2003).
4. A. Bianconi, N.L. Saini, A. Lanzara, M. Missori, T. Rosetti, H. Oyanagi, H. Yamaguchi, K. Oka and T. Ito, Phys. Rev. Lett., **76**, 3412 (1996).
5. E.S. Bozin, S.J.L. Billinge, G.H. Kwei and H. Takagi, Phys. Rev. B, **59**,4445 (1999).
6. B.I. Kochelaev, J. Sichelschmidt, B. Elschner, W. Lemor, and A. Loidl, Phys. Rev. Lett., **79**, 4274 (1997).
7. D. Mihailovic and V.V. Kabanov, Phys. Rev. B, **63**, 054505 (2001).
8. D. Mihailovic, V.V. Kabanov and K.A. Mueller, Europhys. Lett., **57**, 254 (2002).
9. A. Bussman-Holder, K.A. Mueller, R. Micnas, H. Buettner, A. Simon, A.R. Bishop and T. Egami, J. Phys.: Condens. Mat., **13**, L169 (2001).
10. K.A. Mueller, G-M. Zhao, K. Conder and H. Keller, J. Phys.: Condens. Mat., **10**, 291 (1998).
11. A. Lanzara, G-M. Zhao, N.L. Saini, A. Bianconi, K. Conder, H. Keller and K.A. Mueller, J. Phys; Condens. Mat., **11**, L541 (1999).
12. A.W. Hunt, P.M. Singer, K.R. Thurber and T. Imai, Phys. Rev. Lett., **82**, 4300 (1999).
13. D. Rubio Temprano, J. Mesot, S. Janssen, K. Conder, A. Furrer, H. Mutka, and K.A. Mueller, Phys. Rev. Lett., **84**, 1990 (2000).
14. J.G. Bednorz and K.A. Mueller, Adv. Chem. **100**, 757 (1988), Nobel Lecture.

INTRINSIC INHOMOGENEITY AND MULTISCALE FUNCTIONALITY IN TRANSITION METAL OXIDES

A. R. BISHOP
Theoretical Division and Center for Nonlinear Studies
Los Alamos National Laboratory, Los Alamos, NM 87545, USA

Abstract

We briefly review a perspective of transition metal oxides as correlated electron materials governed by functional multiscale complexity. We emphasize several themes: the prevalence of intrinsic complexity realized in the coexistence or competition among broken-symmetry ground states; the origin of landscapes in coupled spin, charge and lattice (orbital) degrees-of-freedom; the importance of co-existing short- and long-range forces; and the importance of multiscale complexity for key material properties, including hierarchies of functional, connected scales, coupled intrinsic inhomogeneities in spin, charge and lattice, consequent intrinsic multiple timescales, and the importance of multifunctional "electro-elastic" materials. Finally, we suggest that such intrinsic multiscale features are characteristic of wide classes of inorganic, organic, and biological matter.

Transition metal oxides (TMOs) have attracted the attention of chemists and physicists for many decades (1) because of the variety of the phenomena they exhibit — particularly a rich spectrum of broken-symmetry ground states, including charge- and spin-density-waves, Jahn-Teller and Peirerls distortions, ferroelectricity and superconductivity. Similarly, each new generation of experimental probes has revealed new aspects of their inherent complexity in spin, charge and lattice (or orbital) degrees-of-freedom, prompting technology (as in the case of ferroelectrics) and challenging theoretical frameworks of the day. One of the heritages for theory and modeling has been an enduring debate between "ab initio" electronic structure calculations and use of reductionist "many-body" Hamiltonians intended to isolate essential collective phenomena. The last decade of advances in high-performance computing has tremendously advanced our ability to "solve" both ab initio and many-body approaches — allowing an appreciation that both are necessary (and neither sufficient alone) but also that the TMOs as a class belong to the rapidly emerging frontier of "Intrinsically Multiscale Systems" (2). This frontier — driven by qualitative advances in available data and experimental resolution and by concomittent advances in modeling strong nonlinearity and complexity, and in computation and visualization — is now common to most disciplines from biology to cosmology. In the field of materials science (structural and electronic) this is equally true, and we are in an exciting era where the demands of technology (e.g. based on nano-science) require new frameworks and concepts with which to understand and predict the multiscale complexity of "strongly correlated materials". TMOs are prominent in this new era because of the richness of the class of materials and of the data available for them. However, they are only members of a huge and growing set of "complex materials", including hard and soft matter: organic, inorganic and biological, crossing artificial boundaries between chemistry, physics and materials science. (3)

The emerging concepts to guide modeling and measuring frameworks for such multiscale complexity are largely yet to be distilled into relative priority or practicality. Nevertheless, we suggest that the following elements need to be included:

- "Strongly correlated" connotes strong coupling between degrees-of-freedom — structural and electronic (magnetic, charge) — which leads to inherent nonlinearities and feedbacks. (4)
- Nonlinearity has itself undergone revolutions of understanding over the last three decades from which we appreciate many typical consequences (5), including: coherence of structures on "mesoscopic" scales (collections of microscopic single particles, atoms, molecules) which control macroscopic observables; and spatio-temporal complexity, including the phenomena of: "landscapes" (6) of metastable/coexistence phases exhibiting global sensitivities to (and therefore turnability by) small/local, internal/external perturbations; glassy (multi-timescale) dynamics with hysteresis, memory, metastability, stochasticity. The landscapes may themselves be dynamic, as in, e.g., spatio-temporal intermittency (typical of many relevant nonlinear equations). An intriguing prospect is that patterns may be adopted transiently for a given function (in response to an internal or external stimulus), and then transitioned to another pattern for a subsequent function. Thus the same system could assume a sequence of structures for a series of functional steps. This appears to be a

modality in biological matter (for example, Ref. 7), efficiently using the same complex system for multiple functions.

- Measurement and modeling techniques which assume periodicity — e.g. local density electronic structure or crystallography — are generally incomplete and can be seriously misleading. At a minimum, they need to be a augmented by specific measurements on important scales (microscopic and mesoscopic). Indeed the validation of new interpretative frameworks for multiscale complexity can only follow from deliberately integrated suites of experiments. Average measurements can help determine parameters in nonlinear models but are rarely answers in themselves.
- Not all scales are equally important. Some carry key functions and it is on these "functional scales" that attention must be focused in building predictive models. (A familiar example is provided by dislocations in solids.) Other scales can then (in principle) be integrated out to create stochastic baths (controlling fluctuations and dissipations) for the functional scales. These intermediate scales are, however, essential because they provide the connections between the functional scales, which act together as a hierarchial system (or "engine"). Thus the spatio-temporal color of the intermediate scales encodes critical information and cannot be approximated arbitrarily (e.g. as white noise). The need for partial methodologies for mesoscopic thermodynamics and statistical mechanics is clear. In short, "hotspots" or "substates (functional scales) are connected in a medium which provides both structural integrity and communication between the hotspots to form the hierarchial system.
- Spatio-temporal complexity is inevitably the consequence of coexisting (and usually competing) lengthscales and, or timescales. (8) This requires careful consideration of adiabatic or slaving approximations in each specific context.
- Since multiscale complexity is intrinsic, we must learn to take advantage of it — and not futilely attempt to avoid it or engineer it away. This means learning to tune relevant properties not only through average properties but also the functional scales and their hierarchial interconnections. In materials, this constitutes a fundamental generalization of the notion of "polarizability" (electronic and structural).
- "Soft" materials are not only those which are geometrically soft or of low density (e.g. polymers). Softness can also connote very deformable unit cell (as in TMOs) or organic macromolecule, and then the constraints on the packing of these deformations in the dense phases composed of many unit cells/molecules leads to mesoscopic patterns. In "hard" materials it has become a traditional approach (mostly of necessity) to identify forces determining structure and then to separately parametrize forces controlling excitations around them. In soft matter this separation may not be possible — the dynamics of the mesoscopic "skeleton" patterns and excitations around those patterns may be intricately coupled — indeed the feedback between function and structure will represent an essential feature of complex matter.
- Temporal aspects of "glassiness" are an automatic consequence of multiple lengthscales. Thus glasses from disorder and frustration (e.g. spin glasses) are only a subset of glasses in which intrinsic scales lead to the same multi-timescale characteristics.
- One large subclass of complex systems is governed by co-existing short-range and long-range interactions, leading to mesoscopic organization and dynamics determined by constraints simultaneously at microscopic (e.g. lattice) scales and through boundaries. Examples include: filamentary flux flow in superconductors (9), charge-ordering in polyelectrolytes (10), clumping in a Van der Waals glass (11), Quantum Hall effect dynamics (12), gels, macromolecular aggregation, and more. The long-range fields maybe explicit (e.g.Coulomb) or the results of local constraints or of topological mesoscopic structures/dislocations, vortices, etc.) which are themselves consequences of local (lattice scale) constraints. As a simple, but already rich example, Fig. 1 shows, the result (13) of a simulation of particles in a 2-dimensional box and interacting via the potential $U(R) = R^{-1} - \exp(-\kappa R)$. As the figure illustrates, varying the relative strengths of the short- and long-range forces leads to patterns of Wigner crystal, filamentary stripes, and clumps. Ref. (13) discusses the fascinating consequences of these patterns on depinning and dynamical ordering in an external field, melting at finite temperatures, competition with pinning from extrinsic disorder, etc.

The basic challenges for research in complex matter are clear: (i) Understanding the microscopic interactions controlling mesoscopic patterns; (ii) Learning how to measure and characterize the properties of the mesoscopic patterns relevant for (specific) macroscopic properties; and (iii) Predicting desired macroscopic properties from (connected hierarchies of) functional mesoscales. This plan is, of course, far more easily stated than achieved. For the remainder of this short discussion, we note some of the recent strategies and progress in the case of TMOs.

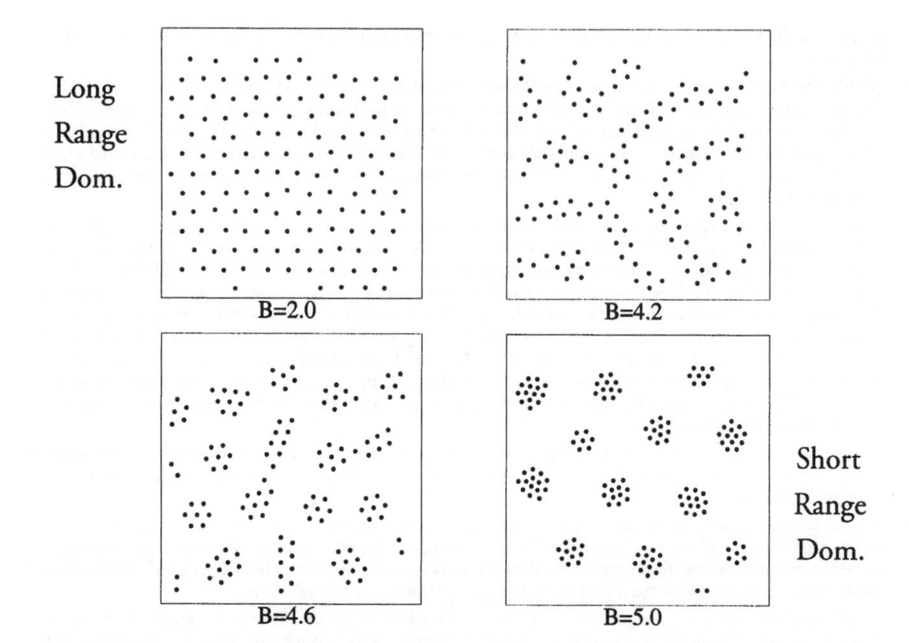

Figure 1 (After Ref.13) Ground states of a 2-dimensional system of particles interacting with combination of short-range attractive and long-range Coulomb repulsive interactions. The systems forms Wigner crystal, stripe and clump phases as the relative strength of the attractive interaction is increased. These phases respond distinctively to applied fields, disorder, thermal melting, etc.

TMOs as a class share the ingredients of highly polarizable oxygen ions together with varying degrees of (dynamic) charge-transfer between the transition metal and oxygen ions. Together, this results in extreme sensitivity to local environment — as witnessed by, e.g., empirical observations of TMO dependence on "oxygen stoichiometry" or by the charge-transfer dependence on lattice vibrations (phonon-assisted charge-transfer) in chemical and biological settings — and probably many solid state TMOs (superconducting cuprates, colossal magnetoresistant manganites, ferroelectrics, etc). An effective phenomenological model which captures these ingredients of TMOs is the "non-linear shell-model". This model is able (14) to describe multiscale coexistence patterns of charge, spin and lattice distortions, as well as delicate features such as isotope-dependences and quantum glass (e.g.quantum paraelectric) phases.

An alternative approach, particularly popular for "correlated electron" many-body modeling, has been to focus on effective orbitals close to the Fermi level (e.g. transition metal d and oxygen p orbitals) and to construct reduced models in those orbitals using selected degrees-of-freedom (e.g. spin or charge), assuming that other degrees-of-freedom can be integrated away and absorbed into renormalization of parameters. Such models (e.g. Hubbard, double-exchange, Kondo) may be expressed in multi-band or effective 1-band forms. The hybridization and charge-transfer (valence fluctuation) is included but typically electron-electron interactions are restricted to short ranges on the assumption of strong itinerant electron screening. Whilst such models can produce inhomogeneous coexistence patterns (15) and may be useful for determining the local (e.g. "stripe") ordering or for excitations (spin, charge, lattice) around those patterns, hierarchies of structure in mesoscopic patterns will more likely be controlled by competitions of short- and long-range forces. (16, 17)

A plausible framework is as follows. The d-orbitals (as for f-orbitals in actinides which exhibit very similar complexities (18)) are both very tightly localized and highly directional. The tight orbitals lead to a narrow electronic band and thus an electronic localization tendency which is enhanced by stage e-e interactions. The localized orbitals can then, locally, couple strongly to globally weak fields, including electron-lattice coupling. This results in strong local lattice distortions, local spin and charge distortions, local charge transfer, local lowering of symmetry — in short, strong (non-thermal) and local variations in chemistry. (19)

Simultaneously, the directionality of localized (non-itinerant) orbitals places strong constraints on bonding orientations from atom to atom — inducing effective long-range coupling, much as bonding constraints lead to landscapes in , e.g., covalent glasses. Thus the patterning of local chemistry (in spin, charge, lattice distortions) on mesoscopic scales is highly constrained into metastable "glassy" configurations. Attempts have been begun to incorporate this competition of short-long-range forces in the cases of doping into broken-symmetry stoichiometric states (AF, charge-density-wave, Jahn-Teller, etc). These early efforts have mostly focused on the short-range arising from lattice-scale distortions induced by the dopant ions in the broken-symmetry, and on the long-range from (poorly screened) Coulomb fields. An example from Ref. 17 is shown is Fig.2. The tendency toward clumps and particularly filamentary ("stripe") mesoscale patterns is strong; consistent with macroscopic onsets of superconductivity, magnetoresistance and ferroelectricity viewed as (dynamic, correlated) percolation phenomena.

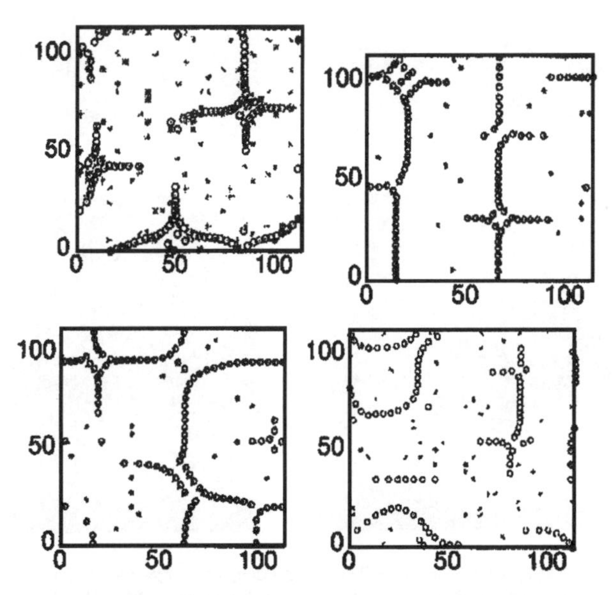

Figure 2. (After Ref. 17) Simulations of mesoscopic patterns in a semiclassical model of a doped CuO2 plane, including short-range fields and long-range Coulomb fields, together with various types of (*) impurities (charged, neutral, in-plane, off-plane). Circles indicate location of dopant charge, and (→) the associated dipole orientation. Note the persistence of fibrillar "stripe segments" in the mesoscopic patterns.

An alternative path to the same conclusion is more compellingly self-consistent. Namely, local (unit cell scale) lattice distortions, constrained by the local bonding directionality, determine not only the local "chemistry" but also intra-inter unit cell (i.e. optic-acoustic phonon) coupling. Specifically, "compatibility" conditions, expressing the constraint of bond bending but not rupture, relate symmetry allowed deformations and directly lead to long-range, directional strain-strain coupling. (20) Thus local constraints lead to long-range fields and hence hierarchial mesoscopic patterns (such as twinning and tweed observed by high-resolution microscopies and neutron/X-ray scattering, etc). Most importantly, local perturbations (specifically from doping) have global consequences as in all landscapes formed from coexisting short-long-range forces. Thus the mesoscopic ordering of dopant charges is strongly driven by elastic fields in a self-consistent interplay of spin, charge and lattice fields: elastic fields are strong. This point of view is developed in Refs 20 and the article of Shenoy et al. in these Proceedings. An example is shown in Fig. (3). An important consequence is that "elastic" and "dielectric" properties in TMOs must be complementary and reflect the multiple scales — indeed there are increasing reports of anomalies in both elastic constants (21) and dynamic polarizabilities (22).

Figure 3. (After Ref. 20 and Shenoy et al., these Proceedings) Coupled charge-strain textures in a Ginzburg-Landau model including local strain-compatibility constraints inducing long-range elastic fields. Left column: strain fields. Right column: charge fields. Note the multiscale patterns — a textured polaron gas above a structural transition temperature To, and twin bands (plus additional microscales) below To.

We note in passing that multiscale complexity and nanoscale phase separation of spin, charge, lattice in complex electronic materials, can also occur without doping away from commensurate stoichiometry. (23) Competition of e-e and e-lattice, multiple bands, and anisotropic coupling are determining in cases such as organic charge-transfer salts or heavy-fermion superconductors. Also, inhomogeneity may be photo- or field-induced in TMOs and other complex electronic materials. For brevity, we do not discuss these cases here, nor competitions of intrinsic with extrinsic (disorder/impurity) mechanisms for inhomogeneity.

Returning to doping-induced multiscale complexity in TMOs, the local distortions are almost certainly in the form of small polarons (e.g. Jahn-teller polarons (24) or generally "textured polarons" (19) in which charge (and/or spin and lattice distortions) are strongly localized in a core (a few unit cells) embedded in a sign-varying strain far-field (above). Then, in high-temperature superconductivity cuprates, T* is the onset temperature for the polaron formulation (the value of T* is dependent on the time-scale of the experimental observation technique). As temperature is further lowered, the multi-polaron configuration stabilizes into the clump and fibrillar mesoscale patterns discussed above — consequences of coexisting, short-long-range fields. There is now much evidence for these mesoscopic polaron patterns and their associated glassy dynamics – see, for example, articles in these Proceedings. The excitations associated with internal (lattice-assisted, local charge-transfer) dynamics of polarons are especially clear (25). There is also strong evidence for novel local vibrations of the lattice (localized phonons) around the edge of polarons or the filaments of clumps. This evidence is now available in cuprates, manganites and nickelates. (26) It clearly correlates with unusual ARPES electronic signatures (kinks), supporting the coupling of electronic and lattice degrees-of-freedom (27). Correlated local (edge mode) signatures in spin and charge with these lattice modes are anticipated and should be sought experimentally. Likewise, low-energy (~meV) signatures of inhomogeneity must correlate with high energy (~eV) ones (electronic gap states); there is some evidence of both modes being important for superconductivity (e.g. 28).

It is worthwhile emphasizing that very little is known about multi-polaron states — progress on single polaron behavior in minimal models does not provide much insight. As the density increases toward filling space, we can anticipate a "quantum" melting to a homogeneous phase with the Fermi surface displaced to the new density – this may be viewed as a "quantum critical point" near optimal doping in HTC cuprates — much as the doped insulator-metal transition in conjugated polymers. (29)

Although there has been some progress in learning how to recognize and quantify mesoscopic inhomogeneities of spin, charge and lattice, the relation to the high-temperature superconductivity mechanism remains tantalizing but unproven. (The alternative viewpoint that inhomogeneity suppresses underlying pairing mechanisms is not disproven by available data.) There is evidence for coexistence of "normal" and superconducting phases below Tc (especially for less than optimal doping). There are also anomalies in, e.g., local structural probes around Tc, persuasive of coupling to the pairing mechanism. However, several scenarios making use of the inhomogeneity for high Tc superconductivity have been proposed and remain plausible without further discriminating data. These include: confinement-induced pairing on the charge-rich regions (30) followed by either (dynamic, correlated) percolation (31) or Josephson coupling (32); and "two-fluid" descriptions (33), including using inhomogeneity-generated low-energy excitations (e.g. the edge-modes above) as bosonic glue for fermions in the charge-poor regions, or coupling magnetic in-plane bands to charge-transfer (incipient ferroelectric) bands perpendicular to the CuO_2 planes which seems to be a clear component of the Jahn-Teller small polarons, supported by structural and optical data (34). Similar temperatures (T^* and Tc) in terms of polaron formation and condensation into percolating mesoscopic networks are implied by data for manganite oxides exhibiting "colossal magnetoresistance".

To summarize, we have briefly surveyed an emerging perspective for TMOs (and other "complex electronic materials") as being intrinsically multiscale. As schmetically illustrated in Fig. (4), this may be viewed as electronically active ("functional") regions (polaronic "hotspots") hierarchically embedded in a self-consistent matrix which provides both structural integrity and essential communication between scales — a self-organized "system" or "engine" much as we can imagine in covalent glasses or biological networks.

More speculative but, if confirmed, extremely powerful is the perspective that the same local intra-unit cell distortions (buckling, coupling of local symmetries such as planar-c-axis in layered perovskites) are responsible for both (a) local, anisotropic "chemistry" (local Jahn-Teller, covalency, dynamic charge-transfer, magnetism, ferroelectric dipoles, etc), and (b) long-range anistropic elasticity. This coexistence of short- and long-range forces then controls the self-organization of hierarchies of patterns in (coupled) lattice, spin and charge — and possibly percolative onsets of macroscopic collective phenomena (magnetism, charge-density, metallicity, superconductivity, ferroelectricity, etc) in multiscale coexisting with other phases.

This point of view defines an exciting era of "electro-elastic" materials, with intrinsically multiscale and multi-functional properties. If we can understand, control and exploit this intrinsic richness, striking new technological flexibility can be envisioned. Already, we can find examples in CMR materials where magneto-resistance and elasto-resistance are fundamentally related (35), or in GMR materials where the structural variation at magnetic interfaces is essential to the device function (e.g.spin-valves), or in cuprate oxides where magnetic "shape-memory" effects are recently observed (as in many perovskie oxides) (36). Essentially the same "electro-elastic" properties govern the functionality of conjugated electronic polymeric materials (37), or of charge-ordering and structural responses (e.g. ion-channels) in biological membranes. Further, the whole field of self-consistently coupling geometry to electronic or magnetic fields is in its infancy.

Transition Metal Oxides: (& other "complex electron materials") An Intrinsic Multiscale Framework

- Charge-spin-lattice coupling essential (orbital)
- Electronically active ("functional") regions (polaronic "hotspots") embedded in a self-consistent hierarchical matrix:

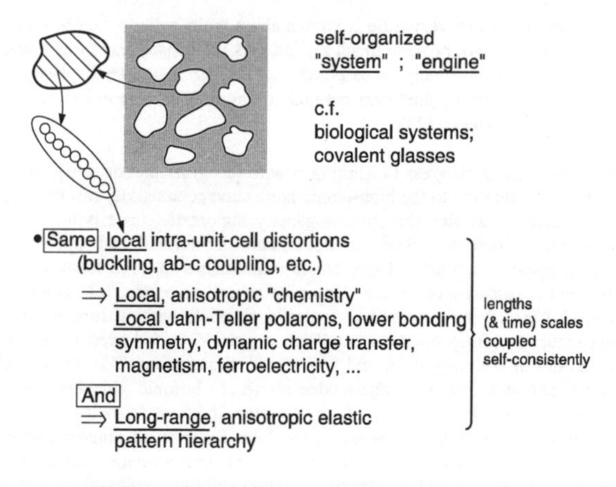

self-organized
"system" ; "engine"
c.f.
biological systems;
covalent glasses
• Same local intra-unit-cell distortions
(buckling, ab-c coupling, etc.)
⇒ Local, anisotropic "chemistry"
Local Jahn-Teller polarons, lower bonding symmetry, dynamic charge transfer, magnetism, ferroelectricity, ...
And
⇒ Long-range, anisotropic elastic pattern hierarchy
lengths
(& time) scales
coupled
self-consistently

- **Different probes detect different parts of the correlated heterogeneity**

Figure 4. A schematic scenario for intrinsic mesoscopic multiscale patterning in transition metal oxides, emphasizing the probable central role of local (intra-unit-cell) distortions for both local chemistry and long-range elastic fields. (See text)

Although the experimental and theoretical protocols for selectively identifying mesoscopic scales are barely available at this time (38), as we begin to acquire many more examples of intrinsic fundamental complexity in "complex electronic materials", the need to seek organizing principles is evident — to provide interpretative frameworks for new generations of experiments and to guide new materials searches. This is imperative as we move beyond academic studies of "single crystals" of inorganic and organic materials to device configurations with thin films, interfaces, and hybrids of inorganic, organic and biological matter. The coupled deformability (elasticity) and electronic polarizability of all these materials will dominate field-injection, photo-injection, spin/charge transport, coherence and decay. In the finite systems of nanoscale science and technology, quantum information, etc. the intrinsic competitions and metastabilities are even more pronounced, just as in biological macromolecules and cellular structures.

A major change of philosophy lies in the need to move beyond the solid state traditions of Bloch states and a linear wave-vector basis. These were themselves a magnificent heritage of quantum mechanics in the last century. However, the assumption of lattice periodicity, sufficient for simpler materials and average properties, fails for the intrinsically inhomogeneous, multiscale patterns of strongly correlated matter — and cannot be ignored if the functionalities reside in those scales. This observation is hardly radical in soft polymeric or biological matter, but in solid state it means we must revise our operational definition of a "single crystal", and apply fresh concepts to notions of e.g. "quantum critical points" or "non-Fermi liquid" behavior. In particular, electronic structure or manybody computational schemes are incomplete if restricted by periodicity assumptions. The lessons are equally important for experimental probes. Thermodynamic, Fermi surface, crystallography, etc. measurements need to be interpreted carefully and provide incomplete information. New generations of probes at local and mesoscopic scales, and with related scales of time resolution are essential. For strongly correlated electron materials, new local probes such as XAFS, neutron PDF, STM are already proving valuable. Seeking elasticity and polarizability signatures and using nonlinear susceptibilities, time-resolved spectroscopies, and noise measurements are directions which need more emphasis. Similarly, correlating the coupled high-energy and low-energy signatures of inhomogeneity are essential. The new concepts we must establish will only be successful if they can explain all these kind of data for spin, charge and lattice degrees-of-freedom, in a common integrating framework. This Workshop represents the beginning of a period of great challenge and promise in which these fundamental issues must be systematically addressed.

I am grateful to many collaborators who have informed my views on this subject, including A. Bussman-Holder, J. Krumhansl, T. Lookman, A. Saxena and S. Shenoy.

REFERENCES

1. e.g., "*Physics of Highly Correlated Electron Systems*", eds. J. O. Wills *et al.*. (North-Holland 1990).
2. e.g., A. R. Bishop, H. Roder, *Current Opinion in Solid State and Materials Science* **2**, 244 (1997); A. R. Bishop, *Synthetic Metals* **86**, 2203 (1997).
3. National Academy Press "*The Physics of Materials; How Science Improves Our Lives*" (1997).
4. e.g., "*Nonlinearity in Materials Science*", Eds. A. Bishop, R. Ecke, J. Gubernaitis (North-Holland 1993).
5. e.g., "*Nonlinearity in Condensed Matter*", Eds. A. R. Bishop *et al.*. (Springer, Berlin, 1987).
6. e.g., H. Westfahl, J. Schmalian, P. G. Wolynes, *Phys. Rev. B* **64**, 174203 (2001) and refs therein.
7. e.g., K. Simons, D. Toomre, *Science* **290**, 1720 (2000).
8. e.g., "*Competing Interactions and Microstructures: Statics and Dynamics*", Eds. R. LeSar *et al.*. (Springer, Berlin, 1988).
9. N.-G. Jensen, D. Dominguez, A. R. Bishop, *Phys. Rev. Lett.* **76**, 2985 (1996).
10. N.–G. Jensen et al.., *Phys. Rev. Lett.* **78**, 2477 (1997).
11. W. Klein *et al..*, *Phys. Rev. Lett.* **85**, 1270 (2001).
12. K. Cooper *et al..*, *Phys. Rev. B* **60**, R11285 (1999).
13. C. Reichhardt, C. Olsen, I. Martin, A. R. Bishop, *EuroPhys.Lett* (2002).
14. See, H. Bilz, G. Benedek, A. Bussman-Holder, *Phys. Rev. B* **35**, 4840 (1987); A. Bussman-Holder *et al..*, *Phil. Mag.* **80**, 1955 (2000) and these Proceedings.
15. e.g., E. Dagotto. T. Holta, A. Moreo, *Phys. Rep.* 344, 1 (2000).
16. e.g., E. Fradkin, S. A. Kivelson, *Phys. Rev. B* **59**, 8065 (1999).
17. B. Stojkovic *et a*, *Phys. Rev. Lett.* **82**, 4679 (1999); *Phys. Rev. B* **62**, 4353 (2000); S. Terber, S. Brazovskii, A. Bishop, *J. Phys. Cond. Matt.* **13**, 4015 (2001).
18. V. A. Sidorov *et al..*, *Phys. Rev. Lett.* (2002) (cond-mat/0202251).
19. K. Yonemitsu *et al..*, *Phys. Rev. B* **47**, 8065, 12059 (1993); Z.-G. Yu *et al..*, *Phys. Rev. B* **57**, R3241 (1998); *J. Phys. Cond. Matt.* **10**, L437 (1998); R. McQueeney *et al..*, *J. Phys. Cond. Matt.* **10**, L437 (1998); R. McQueeney *et al..*, *J. Phys. Cond. Matt.* **12**, L317 (2000).
20. T. Lookman *et al., Phys. Rev. B* (2002); K. Ahn *et al.*, preprint (2002); K. Rasmussen *et al.*, *Phys. Rev. Lett.* **87**, 5704 (2001).
21. e.g., J. L. Sarraro *et al.*, *Phys. Rev. B* **50**, 13125(1994).
22. e.g., S. Sridhar *et al.*, *Phys. Rev. B* **65**, 132501 (2002) and these Proceedings.
23. Y. Yi, A. R. Bishop, H. Roeder, *J. Phys. Cond. Matt.* **11**, 3547 (1999).

24. D. Mihailovic, V. V. Kabanov, *Phys. Rev. B* **63**, 54505 (2001).
25. M. Salkola *et al.*, *Phys. Rev. B* **51**, 8878 (1995); A. R. Bishop, J. Mustre de Leon, D. Mihailovic, preprint (2002).
26. M. Arai et al., Phys. Rev. Lett 69, 359 (1992); R. J. McQueeney et al., Phys. Rev. Lett. 82, 628 (1999); J. M. Tranquada et al, Phys. Rev. Lett. 88, 075505 (2002); T. Egami, these Proceedings.
27. A. Lanzara *et al.*, *Nature* **412**, 510 (2001).
28. W. A. Little, M. J. Holcomb, *J. Supercond.* **13**, 695 (2000).
29. A. J. Heeger, S. Kivelson, J. R. Schrieffer, W. P. Su, *Rev. Mod. Phys.* **60**, 781 (1988).
30. J. Eroles, G. Ortiz, A. Balatsky, A. Bishop, *Eur. Phys. Lett.* **50**, 540 (2000).
31. D. Mihailovic, V. V. Kabanov, K. A. Muller, *Eur. Phys. Lett.* **57**, 254 (2001).
32. I. Martin, G. Ortiz, A. Balatsky, A. Bishop, *Eur. Phys. Lett.* **56**, 849 (2001).
33. See, *e.g.*, A. Bussman-Holder *et al.*, *J. Phys. C* **14**, L165 (2001), and Refs. Therein.
34. See A. Bussman-Holder, these Proceedings.
35. Y. Hwang et al, *Phys. Rev. B* 52, 15046 (1995).
36. A. N. Lavrov, S. Komiya, Y. Ando, *Nature* 418, 385 (2002).
37. S. Tretiak *et al.*, *Phys. Rev. Lett.* (2002); Phase Transitions (2002).
38. J. C. Davis et al., *Nature* **413**, 202 (2001), **415**, 412 (2002) and these Proceedings; A. deLozanne *et al.*, *Phys. Rev. B* **61**, 9665 (2000) and these Proceedings.

MICRO-STRAIN AND SELF ORGANIZATION OF LOCALIZED CHARGES IN COPPER OXIDES

A. BIANCONI, G. CAMPI, S. AGRESTINI, D. DI CASTRO, M.FILIPPI, C.DELL'OMO

Unitá INFM and Dipartimento di Fisica, Università di Roma "La Sapienza"
P.le Aldo Moro 2, 00185 Roma, Italy

It has been recently shown that local lattice distortions in complex oxides are intimately related with the micro-strain in the electronically active structural element, as the CuO_2 plane of doped copper oxide superconductors. In fact, the micro-strain of the Cu-O bond controls the electron-lattice interaction. Here we have investigated charge ordering in an oxygen doped $La_2CuO_{4.1}$ crystal by high-resolution x-ray diffraction using synchrotron radiation. Thanks to the high brilliance synchrotron radiation it has been possible to record a large number of weak superstructure spots due to charge ordering around the main peaks of the average structure. A study of the charge modulations with stage 3.5, and their behavior as a function of the intensity of the x-ray incident flux is reported. Using an approach based on joint x-ray photo-excitation and x-ray diffraction we are able to see the self organization of photo-doped charges into short range polaron striped bubbles (a 2D superstructure q_3) and a 3D polaron crystal of strings (a 3D superstructure q_4).

1. Introduction

The doped cuprates provide a unique superconducting "condensate", that is robust enough to sustain temperature up to T_c~150 K. The non-Fermi liquid behavior of cuprates has been known, since 1987, from the linear temperature dependence of the electrical resistivity extending up to very high temperature. Recently experiments providing evidence for nearly critical fluctuations in the lattice [1,2], spin [3] and charge [4] channel have been reported. However the phase diagram of the normal phase remains a point of unresolved debate. It has been known that the chemical pressure [5-10] (the compressive shear stress on the bcc CuO_2 layer due to lattice mismatch with the fcc rock-salt layers) induces variations in the critical temperature T_c. Indeed recent experiments suggest that the physical properties of the cuprates depend systematically on the micro-strain ε of the Cu-O bond [11]. The micro-strain of the CuO_2 plane $\varepsilon = 2\frac{r_0 - \langle r_{CuO}\rangle}{r_0}$ (where r_0=1.97 Å is the equilibrium Cu-O distance with no chemical pressure) is determined by a direct measurement of the in plane Cu-O bond $\langle r_{CuO}\rangle$ by a local probe, the Cu K-edge extended x-ray absorption fine structure (EXAFS).

The recently reported three-dimensional (3D) phase diagram for hole doped cuprate perovskites as a function of doping, strain and temperature is shown in Fig. 1. The critical temperature as a function of micro-strain and doping is plotted (lower). The micro-strain varies from one family to the other and/or by atomic substitution in the rock-salt sublattice. The critical temperature T_c reaches a maximum (~130K) at a critical micro-strain (ε_c~0.04) and in the doping range δ~0.16±0.03. Different experiments in the normal state of the cuprates can be schematically classified according to the different regions of the 3D phase diagram shown in Fig. 1 (upper). Electronic striped crystals with long range order appear at very large micro-strain ($\varepsilon > \varepsilon_0$~0.075), and for commensurate doping (1/8). Bubbles of about 100-300 Å diameter of fluctuating stripes with wave length of about 14 Å show up in the range $\varepsilon_c < \varepsilon < \varepsilon_0$. The superconducting phase appears in the range $\varepsilon_c/2 < \varepsilon < \varepsilon_0$. Assuming that the micro-strain controls the electron lattice coupling [12,13], these findings are consistent with the prediction of a critical electron lattice coupling for the onset of local lattice distortions at the metallic densities [14,15]. The superconductivity occurs in a particular inhomogeneous electronic phase [16-18] as described for the doped perovskites [19] and magnetic semiconductors [20].

Considering the plane of temperature and micro-strain, at constant optimum doping, the superconducting phase below T_c appears for the range $\varepsilon_c/2<\varepsilon$ as shown in Fig. 2. For the intermediate micro-strain $\varepsilon_c<\varepsilon$ there is a temperature range T_{sf} for the stripe formation [18] (shaded area) that separate the fluctuating stripe bubbles at low temperature from the polaron liquid at high temperature. In the range of micro-strain where superconductivity appears, quantum lattice fluctuations are expected [21]; in fact, the micro-strain is expected to control the strength of the electron-lattice interaction as also indicated by recent studies on isotope effect [22-24].

The oxygen doped $La_2CuO_{4+\delta}$ superconducting system, having a high value of micro-strain, ε=0.074, is the simplest system to study the ordering of dopants, and the charge ordering in the CuO_2 plane of high T_c superconductors. The mobile oxygen as dopants makes it a very special case in the families of La-based cuprates high T_c superconductors [25-28]. Recently we have shown formation of stripe domains [28, 29] in the $La_2CuO_{4+\delta}$ system and their possible consequences on the characteristic properties of this system. In this paper we report influence of temperature and x-ray illumination on the inhomogeneous phase of oxygen doped $La_2CuO_{4+\delta}$ (δ~0.1), by high-resolution X-ray diffraction measurements.

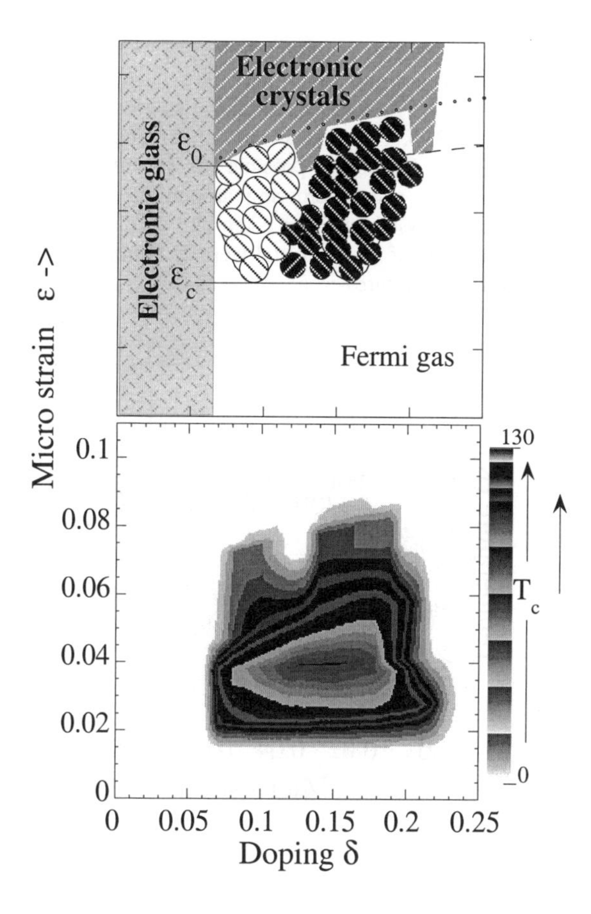

Figure 1. The phase diagram of all doped cuprate superconductors: The contour plot of superconducting temperature T_c as a function of micro-strain and doping (lower). A cartoon picture of the phase diagram at T=0 (upper); normal phase A(normal metal), B(bubbles of metallic stripes) C (long range ordered striped crystal) D (electron glass).

2. Experimental details

Diffraction measurements on the $La_2CuO_{4.1}$ single crystal, grown by flux method and doped by electrochemical process [30], were performed on the crystallography beam-line at the Elettra storage ring at Trieste. The X-ray beam emitted by the wiggler source on the Elettra 2 GeV electron storage ring, was monochromatized by a Si(111) double crystal monochromator, and focused on the sample. The temperature of the crystal was monitored with an accuracy of ±1K. We have collected the data in the K geometry, with a photon energy of 12.4 KeV (wavelength λ=1Å), using a CCD detector assembly. The sample oscillation around the **b** axis was in a range 0<θ<30°, where θ is the angle between the direction of the photon beam and the **a** axis. We have investigate a portion of the reciprocal space up to 0.6 $Å^{-1}$ momentum transfer, i.e., recording the diffraction spots up to the maximum indexes 3, 3, 19 in the **a***, **b***, **c*** direction respectively. Thanks to the high brilliance source, it has been possible to record a large number of weak superstructure spots due to charge ordering around the main peaks of the average structure. Twinning of the crystal has been taken into account to index the superstructure peaks. The lattice parameters of single crystal were determined to be a=5.351 Å, b=5.418 Å, c=13.171 Å at room temperature.

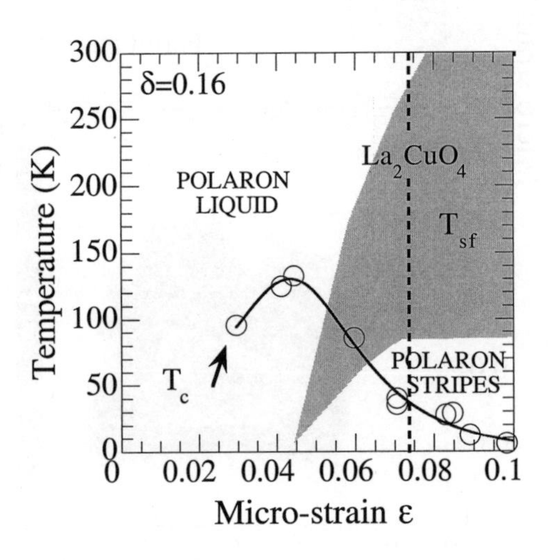

Figure 2. The 2D plane (temperature versus micro-strain at constant doping δ=0.16) cutting the 3D phase diagram, showing the T_c, and the shaded range of the stripe formation T_{sf}.

3. Results and discussions

At room temperature we have observed four types of superstructure peaks [28]. The superstructures are classified according to the staging of oxygen [9], i.e., the modulation wave-vectors along the **c** axis. The first two superstructures are characterized by narrow, resolution limited diffraction peaks showing 3D ordering with the wave-vectors (stage 2): $\mathbf{q_1}$ = 0.089 (±0.003)**a***+0.248 (±0.002)**b***+0.495 (±0.005)**c***; and $\mathbf{q_2}$ = 0.049 (±0.003)**a***+0.268 (±0.002)**b***+0.490 (±0.005)**c***. The modulation with wavevector $\mathbf{q_1}$ is commensurate with the lattice and shows the formation of a crystal made of charge strings of finite length of about 11a, separated by d = 4b and c-axis unit cell doubling [28, 29]. On the other hand the $\mathbf{q_2}$ is an incommensurate superstructure.

The superstructure peaks with $\mathbf{q_1}$ and $\mathbf{q_2}$ coexist with other diffuse spot [28] with the wavevector (stage 3.5): $\mathbf{q_3}$ = 0.208 (±0.003)**b***+0.290 (±0.005)**c*** and an overlapping narrower (resolution limited) diffraction spot, with wavevector (stage 3.5) $\mathbf{q_4}$ = 0.037 (±0.001)**a***+0.198 (±0.002)**b***+0.290(±0.005)**c***. It should be recalled that the $\mathbf{q_3}$ modulation appears with negligible **a*** component and it is diffused along this direction. Therefore the $\mathbf{q_3}$ is assigned to be due to formation of 2D and short range ordered domains (O3 phase). The $\mathbf{q_4}$, on the other hand, is resolution limited superstructure due to the formation of 3D and long range ordered domains (O4 phase). In the phase O3 long stripes (145 lattice units i.e., ~ 77 nm long in the **a**-direction), form an incommensurate superlattice with a period of 4.8 lattice units in the **b** direction. The O4 domains are characterized by a crystal of strings of finite length, (27 lattice units, i.e., ~ 15 nm long in the **a** -direction), forming a commensurate superlattice with a period of 5 lattice units in the **b** direction (~ 27 nm) [28. 29]. Here we focus only on the stage 3.5 superstructures, $\mathbf{q_3}$ and $\mathbf{q_4}$, associated with the two charge ordered phases, called O3 and O4 respectively. The stage 2 superstructures, $\mathbf{q_1}$ and $\mathbf{q_2}$, are discussed elsewhere [28].

Our experimental strategy consists on looking at the effect of photo-doping on the two superstructures $\mathbf{q_3}$ and $\mathbf{q_4}$ at different temperatures. It should be mentioned that the similar approach was earlier applied [31] to study order to disorder like transition by continuous x-ray illumination, in such a way to create photo doped charges in a surface layer of thickness H, determined by the x-ray penetration depth, using high intensity x-ray flux. This approach makes it possible to probe the charge ordering in the same slab by x-ray diffuse scattering with the use of fast 2D detection. In this process of photo-illumination, the electrons are ejected from the CuO_2 plane, leaving an itinerant hole to the insulating block layers. Therefore the use of x-ray illumination adds itinerant holes in the CuO_2 plane and electrons in the LaO planes, and hence their ordering could be studied by the x-ray diffraction. In particular, in order to study the behavior of photo-doped charges we have explored the effect of illumination dose at the three different temperatures 100, 220 and 300K.

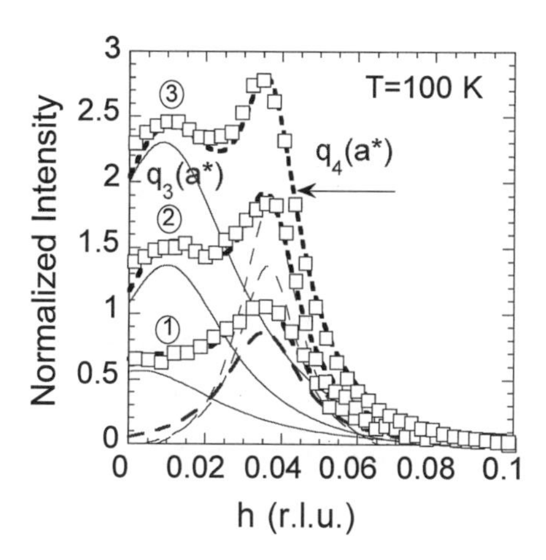

Figure 3. Profiles of the superstructures q_3 and q_4 in the Q= (0,h,6+0.29) direction for three different dose values (denoted by 1, 2 and 3) at T=100K.

The sample was cooled down to T=100K from the highest temperature to investigate the effect of the x-ray illumination at low temperature, T<180K, where the EXAFS data show the existence of a polaron formation. We have observed an increase of the charge modulation, indicated by the behaviour of the intensity of diffraction satellites due to the short range, diffuse, superstructure $\mathbf{q_3}$, and to the resolution limited superstructure $\mathbf{q_4}$, revealing the short range polaron ordering of the photo-doped charges. Fig. 3 shows the profiles of the two superstructures at representative x-ray dose values indicating a clear photo-induced ordering.

Fig. 4 shows evolution of the relative weight of the superstructures $\mathbf{q_3}$ and $\mathbf{q_4}$ as a function of x-ray doses. Here the relative weight corresponds to the integrated intensity I/I_0, where I is the intensity of the peak under illumination while I_0 without illumination. We observe that weight of the peak $\mathbf{q_3}$ and $\mathbf{q_4}$ is increased by ~ 2.5 and 2.0 times respectively of its initial value. The relative weight of the two ordered phases change as under x-ray illumination, as evident also from the Fig. 3 showing the profiles of the two superstructures.

Lets now move to see what happens to the ordering of the photo-induced charges at another representative temperature, ~220 K. Fig.5 shows diffraction profiles of the superstructures $\mathbf{q_3}$ and $\mathbf{q_4}$, measured at this temperature, for different dose values (denoted by 1, 2 and 3). While the superstructures $\mathbf{q_3}$ shows a substantial change in its intensity with increasing dose values of x-ray photons, the $\mathbf{q_4}$ does not seem to be influenced. At 220 K, the peak $\mathbf{q_3}$ shows an increase by ~ 2 times of its initial value, indicating an incommensurate ordering of x-ray photo induced charges in the striped domains (phase O3). Indeed the phase O3 springs up under the x-ray illumination as shown in the Fig. 6, in which an evolution of the normalized integrated intensity I/I_0, is plotted.

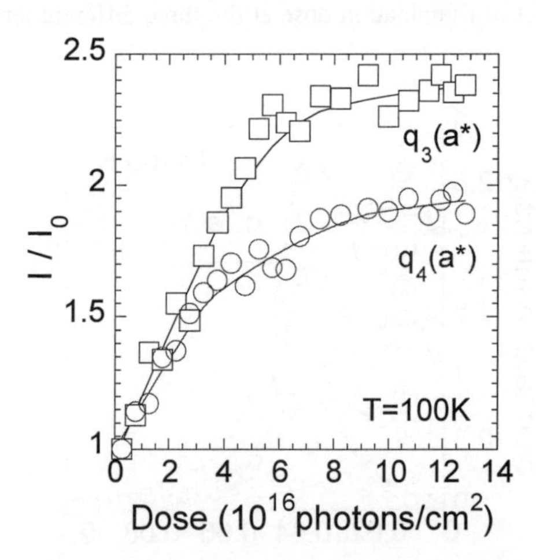

Figure 4. Effect of the x-ray illuminationon the integrated intensity of the charge ordering modulations q_3 and q_4 at T=100 K where I_0 is the initial value of the integrated intensities, without the illumination effect. The numbers 1, 2 and 3 correspond to the dose values indicated in figure 3.

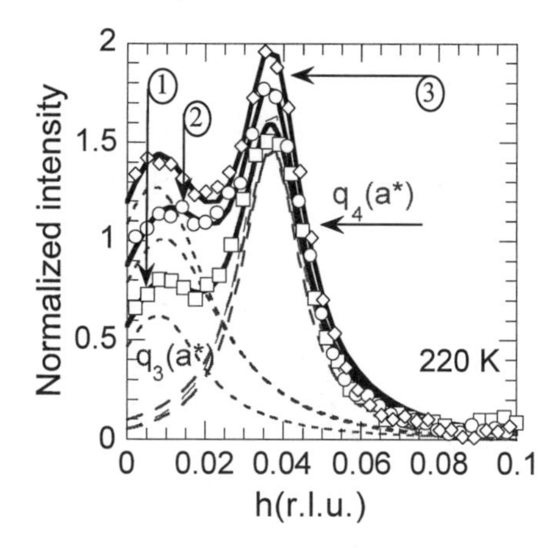

Figure 5. Profiles of the superstructures q_3 and q_4 in the Q= (0,h,6+0.29) direction for three different dose values (denoted by 1, 2 and 3) at T=220K.

Finally, the sample was studied under x-ray illumination at room temperature T=300K. At this temperature we expected to find no relevant illumination effect on the charge ordering, since the point (T=300K, ε=0.074), in the phase diagram, fall out the shadow region (see e.g., Fig.1,). Indeed this is the temperature regime in which the charges start to get arranged themselves into the ordered polaron domains, (associate with $\mathbf{q_3}$ and $\mathbf{q_4}$ superstructures). Fig. 7 reveals consistent behaviour of integrated intensities (I/I_0) of $\mathbf{q_3}$ and $\mathbf{q_4}$ peaks as a function of the illumination at 300K. In fact, the two superstructures appear unaffected by the x-ray pumping, i.e., the photo-doped charges don't seem to have a tendency to self-organize in ordered domains at this temperature.

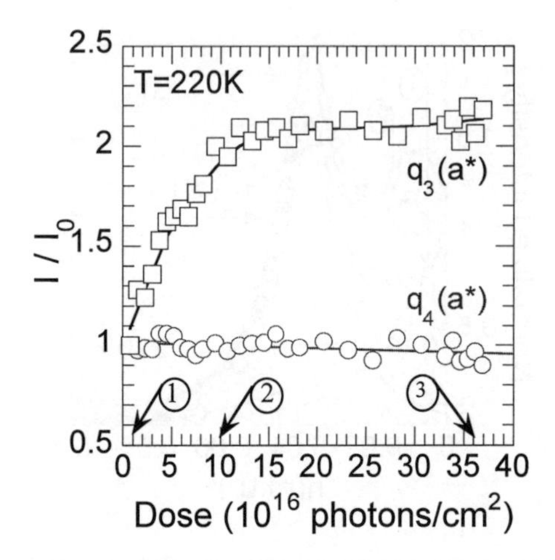

Figure 6. Effect of the x-ray illumination on the integrated intensity of the charge ordering modulations q_3 and q_4 at T=220 K where I_0 is the initial value of the integrated intensities, without the illumination effect. The photoping induces a variation only of the supestrucrture q_3.

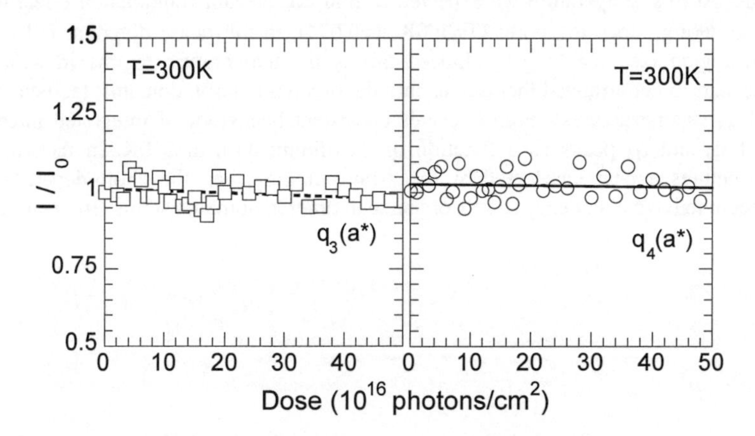

Figure 7. Effect of the x-ray illumination on the integrated intensity of the charge ordering modulations q_3 and q_4 at T=300 K where I_0 is the initial value of the integrated intensities, without the illumination effect. The photoping does not induce any variation on the supestrucrtures q_3. and q_4.

In summary we have studied effect of temperature and x-ray illumination on the superstructures in oxygen doped $La_2CuO_{4.1}$ single crystal by x-ray diffraction measurements. The experimental data shows coexistence of different phases in the $La_2CuO_{4.1}$ sample. The charge modulations with stage 3.5, due to ordered domains of the local lattice distortions, are associated to the superstructure peaks characterising crystal of strings (phase O4) and a superlattice of stripes (phase O3). The study of x-ray illumination at 100, 200 and 300 K, allowed to observe different temperatures below which the x-ray induced charges are able to get ordered in the phase O3 and O4 respectively. The phase O4 appears quite stable, remaining nearly unchanged either under x-ray illumination at 220 and 300K, while the O3 phase is nearly stable only at 300K. At low temperatures, T=100 K, the photo-doped charges self-organize in both O3 and O4 phases. In conclusion, this study provides us a mean to confirm and update the phase diagram for the cuprate oxides, in the case of a material, $La_2CuO_{4.1}$, with high micro-strain. Furthermore, the experimental approach reported in this paper provides possibilities to manipulate the charge ordered phases by temperature and x-ray illumination in complex systems as the high T_c cuprates.

Acknowledgements

This work is supported by "progetto cofinanziamento Leghe e composti intermetallici: stabilità termodinamica, proprietà fisiche e reattività" of MURST, by Istituto Nazionale Fisica della Materia (INFM), and by "Progetto 5% Superconduttività" of Consiglio Nazionale delle Ricerche (CNR).

References

1. R. P. Sharma, S.B. Ogale, Z.H. Zhang, J.R. Liu, W.K. Wu, B. Veal, A. Paulikas, H. Zhang and T. Venkatesan, *Nature* **404**, 736 (2000) and references therein.
2. E. S. Bozin, G. H. Kwei, H. Takagi, and S. J. L. Billinge, *Phys. Rev. Lett.* **84**, 5856 (2000) and references therein.
3. G. Aeppli, T. E. Manson, S. M. Hayden, H. A. Mook and J. Kulda *Science* **278**, 1432 (1997).
4. T. Valla, A. V. Fedorov, P. D. Johnson, B. O. Wells, S. L. Hulbert, Q. Li, G. D. Gu, and N. Koshizuka *Science* **285**, 2110 (1999).
5. J.B Goodenough,. and J.-S. Zhou, *Chemical Materials* **10**, 2980 (1998); also see e.g. C. N. R. Rao and A. K. Ganguli *Chem. Soc. Rev.* **24**, 1 (1995).
6. J.P. Attfield, A.L. Kharlanov and J.A. McAllister, *Nature* **394**, 157 (1998).
7. J.P. Locquet, J. Perret, J. Fompeyrine, E. Mächler, J.W. Seo and G. Van Tendelloo, *Nature* **394**, 453 (1998).
8. Y. Cao, T. L. Hudson, Y. S. Wang, S. H. Xu, Y. Y. Xue, and C. W. Chu. *Phys. Rev. B* **58**, 11201 (1998) and references therein.
9. H. Sato, A. Tsukada, M. Naito, A. Matsuda, *Phys. Rev.* B **61**, 12447 (2000).
10. M. Marezio and F. Licci, *Physica* C **282** 53(1997).
11. A. Bianconi, G. Bianconi, S. Caprara, D. Di Castro, H. Oyanagi and N.L. Saini, *J. Phys. Cond.Mat.* **12**, 10655 (2000).
12. S. R. Shenoy, V. Subrahmanyam, A. R. Bishop, *Phys. Rev. Lett.* **79**, 4657 (1997).
13. F. V. Kusmartsev, D. Di Castro, G. Bianconi, A. Bianconi *Physics Letters* A **275**, 118-123 (2000).

14. S. Caprara, M. Sulpizi, A. Bianconi, C. Di Castro, and M. Grilli, *Phys. Rev. B* **59**, 14980 (1999) and references cited therein; S. Ciuchi and F. De Pasquale, *Phys. Rev. B* **59**, 5431 (1999)
15. A. Blawid and A. Millis, *Phys. Rev. B* **62**, 2424 (2000) and references therein.
16. A. Bianconi, N.L. Saini, A. Lanzara, M.Missori, T. Rossetti, H. Oyanagi, H. Yamaguchi, K. Oka and T. Ito, *Phys. Rev. Lett.* **76**, 3412 (1996) and references therein
17. K. A. Müller, Guo-meng Zhao, K. Conder, and H. Keller *J. Phys. Condens. Matter* **10** L291 (1998).
18. J. B. Goodenough and J. S. Zhou, *Nature* **386**, 229 (1997).
19. S. Mori, C.H. Chen, S.-W. Cheong, *Nature* 392 473 (1998) and references therein.
20. E. L. Nagaev, *Physics of Magnetic Semiconductors* Mir Publisher, Moscow, 1983.
21. S. Sachdev *Quantum Phase Transitions* Cambridge Univ. Press, New York, (1999)
22. J. Hofer, K. Conder, T. Sasagawa, Guo-meng Zhao, M. Willemin, H. Keller, and K. Kishio, *Phys. Rev. Lett.* **84,** 4192 (2000) and references therein.
23. A. Lanzara, G.-m. Zhao, N.L. Saini, A. Bianconi, K. Conder, H. Keller and K.A. Müller, *J. Phys.:Condens. Matter* **11** L541 (1999).
24. D. Rubio Temprano, J. Mesot, S. Janssen, K. Conder, A. Furrer, H. Mutka, and K. A. Müller, *Phys. Rev. Lett.* **84**, 1990 (2000).
25. B. O. Wells, Y. S. Lee, M. A. Kastner, R. J. Christianson, R. J. Birgeneau, K. Yamada, Y. Endoh, and G. Shirane, *Science* **277**, 1067 (1997).
26. J. C. Grenier, N. Lagueyte, A. Wattiaux, J. P. Doumerc, P. Dordor, J. Etourneau, M. Pouchard, J. B. Goodenough and J. S. Zhou, *Physica C* **202**, 209 (1992)
27. C. Chaillout, J. Chenavas, S. W. Cheong, Z. Fisk, M. Marezio, B. Morosin and J. E. Schirber, *Physica C* **170**, 87 (1990).
28. A. Bianconi, D. Di Castro, G. Bianconi, A. Pifferi, N.L. Saini, F.C. Chou, D.C. Johnston and M. Colapietro *Physica C* **341-348**, 1719-1722 (2000); D. Di Castro, M. Colapietro and G. Bianconi, *Int. J. Mod. Phys.* **B14**, 3438 (2000); D. Di Castro et al, to be published (2002).
29. F. V. Kusmartsev, D. Di Castro, G. Bianconi, A. Bianconi, *Physics Letters* A **275** (No. 1-2), 118-123 (2000).
30. F. C. Chou, D. C. Johnston, S. W. Cheong and P. C. Canfield, *Physica* **C216**, 66 (1993).
31. G. Bianconi, D. Di Castro, N.L. Saini, A. Bianconi, M. Colapietro, A. Pifferi, (2000) in "*X-ray and Inner Shell Processes*" eds by R.W. Dunford et al. (AIP proceedings, Woodbury, NY) pag. 358-371.

STRAIN, NANO-PHASE SEPARATION, MULTI-SCALE STRUCTURES AND FUNCTION OF ADVANCED MATERIALS

S. J. L. BILLINGE
Dept. Physics and Astronomy
Michigan State University
4263 Biomedical Physical Sciences Building,
East Lansing, MI 48824, USA
E-mail: billinge@pa.msu.edu

Recent atomic pair distribution function results from our group from manganites and cuprate systems are reviewed in light of the presence of multi-scale structures. These structures have a profound effect on the material properties.

1. Introduction

In advanced materials with interesting functionality a realization is growing that the properties depend sensitively on complex structures on different length-scales from atomic to macroscopic.[1,2,3,4,5] The challenge to experimentalists is properly to characterize such materials. This will be a prerequisite to obtaining a complete theoretical understanding of these materials. It is particularly difficult because many of the existing technologies for studying a material's structure quantitatively reveal only average properties such as the crystal structure. However, key information about higher level structures, potentially driving the interesting properties, is contained in deviations from this average picture. New approaches to studying structure are therefore required to solve this problem. A number of imaging techniques now exist such as TEM, STM, AFM and so on and these are proving to be extremely important.[6,7] What is needed in addition is bulk probes that provide quantitative information about atomic structures in disordered systems. XAFS is an important technique for very short-range structure.[8] Here we describe insights that we have gained from using the atomic pair distribution function (PDF) analysis of powder neutron and x-ray diffraction data.[9]

This technique is described in detail elsewhere.[9] Here we mention only that it yields quantitative information about atomic structure on short (nearest neighbor) and intermediate (up to $\sim$ 10 nm) length-scales. The PDF is obtained by a sine Fourier transform of properly corrected and normalized x-ray or neutron powder diffraction data:[9]

$$G(r) = 4\pi r[\rho(r) - \rho_0] = \frac{2}{\pi}\int_0^\infty Q[S(Q) - 1]\sin Qr\, dQ, \qquad (1)$$

where $\rho(r)$ is the microscopic pair density, ρ_0 is the average number density, $S(Q)$ is total structure function that is the normalized scattering intensity, and Q is the magnitude of the scattering vector, $Q = |\mathbf{k} - \mathbf{k_0}|$. For elastic scattering, $Q = 4\pi \sin\theta/\lambda$, where 2θ is the scattering angle and λ is the wavelength of the scattering radiation.

New insights have been gained into the complex oxides by applying this approach. In particular, here we give an overview of results from cuprate superconductors and colossal magnetoresistant manganites. The role (or otherwise) of the structure in the properties of these materials has been hotly debated over the years. The strong electron-electron interactions and resulting magnetism clearly are important; however, different views are held about the fundamental importance or otherwise of the lattice to the superconductivity and magnetoresistance of these materials. Although phonons were dismissed early as the sole mediators of pairing in the cuprates, a number of studies suggested subtle lattice effects occurring.[10,11] though contradictions and disagreements between the results from different techniques made these hard to assess and interpret. In the manganites, because of the active Jahn-Teller distortion, the magnitudes of the structural effects are much larger and there is much better agreement between the techniques in this case. In fact, lessons learned from manganite research is resulting in renewed efforts and better understanding of the situation in the much more subtle cuprates. For this reason, we begin with the manganites. This is not an extensive review of the field of lattice effects in the cuprates, but an overview of the discoveries that our group has made using the PDF technique.

2. Manganites

2.1. *Polarons and colossal magnetoresistance*

As early as the 1950's the general relationship between magnetism and charge transport was elucidated through Zener's double-exchange mechanism.[12] The important role of the Jahn-Teller effect was noted early on by Goodenough,[13] but its importance to the CMR phenomenon was only properly elucidated in the mid 1990's by Millis *et al.*,[14] and Bishop and co-workers.[15] The PDF provided some of the best experimental evidence that charges were localizing as lattice (and spin, but the PDF only sees the lattice) polarons at the metal-insulator transition, T_{mi}.[16] The evidence appeared in the PDF as an anomalous broadening of the PDF peaks associated with T_{mi}. The nature of the polaronic state was also studied at this time by modelling the structural changes evident in the PDF at T_{mi}.[16,17] The positive charge carriers (holes) localize on Mn^{4+} sites with a shrinking of the MnO_6 octahedron to a small, regular octahedral shape. The octahedra without a localized charge take on a Jahn-Teller elongated shape with a particularly long bond of $r = 2.15$ Å, compared with $r = 1.91 - 1.96$ Å for the short Mn-O bonds present.[17] The appearance of polarons in the insulating phase is unambiguous in the local structure measurements.[16,17,18,19,20]

2.2. *FM phase is nanophase-segregated*

The next phase was to understand how these objects appeared. Do the octahedra uniformly grow with increasing temperature as T_{mi} is approached or do fully distorted polarons locally appear and grow in number? A high resolution x-ray PDF measurement,[17] strongly suggested the latter. In agreement with supporting neutron measurements it seemed that long bonds always appear at the distance expected for unstrained JT distorted octahedra (as seen in undoped $LaMnO_3$ for example) and the number of such states increases with temperature towards T_{mi}. This is shown in Fig. 1.

Interestingly, a significant number of JT distorted octahedra are evident well below T_{mi} in the ferromagnetic metallic (FM) phase. Localized electronic states are coexisting with delocalized metallic states in this region. This suggests a phase separation into conducting and insulation regions and an inhomogeneous electronic structure. Similar behavior was directly observed from dark field imaging TEM studies of a related manganite.[6] However, in that case the insulating and metallic phases were micron sized and static. In this case of the high-T_c $La_{1-x}Ca_xMnO_3$ system *there is no diffraction or TEM evidence of macroscopic phase separation.* Nonetheless, we are forced to the conclusion from the PDF results that the samples are phase separated on a nanometer length-scale, either statically or dynamically, and that this nanostructure evolves with temperature as the phase transition is approached. This general picture in this, and related, manganites is now supported by range of experimental evidences.[21]

2.3. *Percolation transition*

This picture suggests that the MI transition may occur by a percolation mechanism so this was investigated. When the JT distorted octahedra

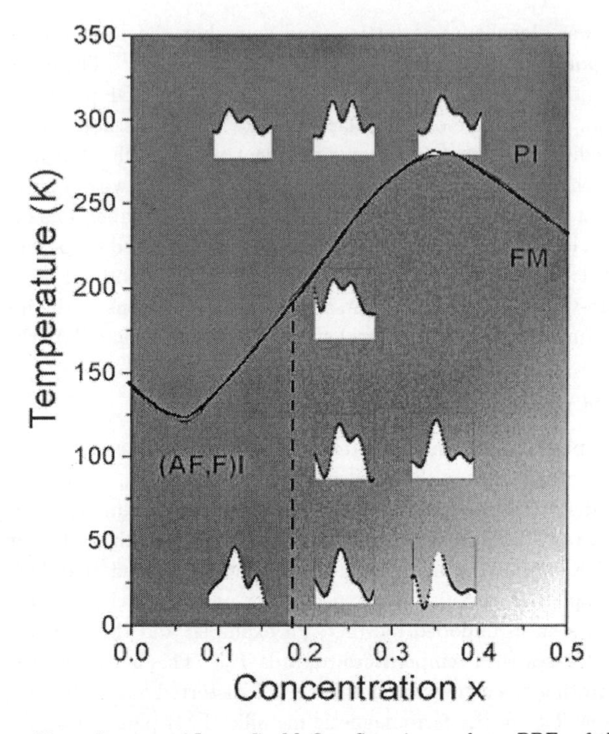

Figure 1. Phase diagram of $La_{1-x}Ca_xMnO_3$. Superimposed are PDFs of the low-r region measured using high-energy x-rays. The peak at lower-r is at the position of the short Mn-O bonds in an MnO_6 octahedron, the higher-r peaks in the doublets are at the position of the Mn-O long-bonds in JT distorted octahedra. Clearly JT distorted octahedra are present throughout the FM region of the phase diagram, except at high doping ($x > 0.25$) and low-temperature. They grow in number as T_{mi} is approached.

appear the PDF peak at 2.75 Å, coming from the O-O bonds on the MnO_6octahedron, broadens.[16] Since the PDF is a bulk measurement, this gives us a semi-quantitative measure of how many octahedra are in the metallic phase. The peak width is inversely proportional to the peak height, hence $h \propto \left(\frac{1}{\langle u^2 \rangle}\right)^{0.5}$. As the sample goes into the metallic phase polarons disappear, $\langle u^2 \rangle$ decreases and the peak-height increases. We were able to show that this behavior scales with reduced temperature, as shown in Fig. 2. These changes in peak height are really coming from the loss of polaronic behavior. In the T-dependent studies we had to subtract the normal T-dependence (the Debye-Waller behavior) before revealing the peak-height scaling shown in Fig. 2. Recently, we have measured a particular isotopically substituted (^{18}O for ^{16}O) sample where the transition into the FM phase is suppressed by the isotope substitution. The results are shown in Fig. 3. The abrupt increase in PDF peak-height on entering the metallic phase is unambiguous.

It is dangerous to call this an order parameter since this tends to imply second order behavior that is not evident here. However, it is a direct experimental measure of the degree to which a sample is in the metallic phase. It can be compared to mixing parameters derived from transport measurements[22] and used in percolative models for transport.[23] Despite it not being an order parameter, Fig. 2 suggests that it shows universal-like behavior. This is also supported by analysis of data from an $La_{1-x}Ca_xMnO_3$ $x = 0.5$ sample described below. This behavior seems interesting and begs a theoretical explanation. Note also that no macroscopic phase separation has been observed in the $La_{1-x}Ca_xMnO_3$ system described here and by Jaime *et al.*[22] and Mayr *et al.*.[23] This places the current system somewhere between conventional notions of first and second order behaviors with apparently both kinds of behavior being observed depending on the length-scale and time-scale of the measurement; a classic multi-scale problem reminiscent of discussions about the phase transitions of $BaTiO_3$.[24] Actually, in the $La_{1-x}Ca_xMnO_3$ system a crossover from first order to second order behavior has been postulated from specific heat measurements as a function of doping, with first order behavior noted below $x < 0.4$.[25]

The problem with the PDF peak-height parameter is that it doesn't yield the absolute value of the metallic fraction but rather only tracks the *change* of this parameter with temperature (or doping, for example). We have attempted to extract quantitatively the absolute fraction of undistorted octahedra by modelling the PDF. Models for the undistorted and distorted phases are taken from the $La_{1-x}Ca_xMnO_3$ $x = 0.3$ and $x = 0.0$ samples, respectively, at low temperature. We are interested in the proportion of distorted and undistorted octahedra rather than how they arrange in space so the PDF is fit only over a range to ~ 4 Å. A characteristic fit is shown in Figs. 4(a) and (b) and the resulting phase fractions are shown in Fig. 4(c). These indicate that this sample ($x = 0.25$) is more than 95% in the metallic phase at low temperature but, surprisingly, remains at least 50% undistorted above T_{mi}. This appears to be supported by neutron small angle scattering measurements which see significant nanometer-scale ferromagnetic clusters surviving far above T_{mi}.[26] A word of caution is necessary here. The phase fractions determined from the PDF fitting will depend on the structural models used for the two phases. A more appropriate model

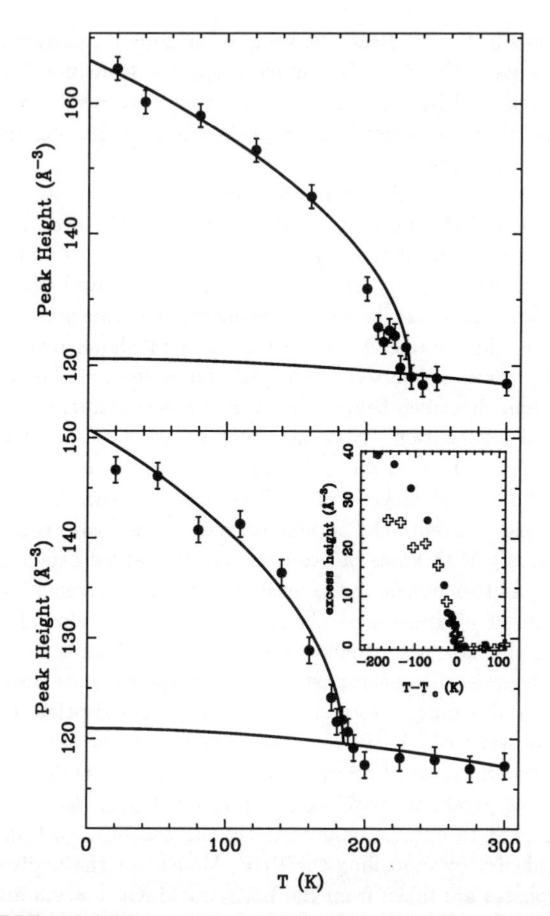

Figure 2. PDF peak height vs. temperature for the peak in $La_{1-x}Ca_xMnO_3$ at $r = 0.275$ that originates from O-O correlations on the MnO_6 octahedra. The panels show the $x = 0.25$ (top) and $x = 0.21$ (bottom) compositions. The lower solid line is the expected Debye behavior fit to the high-T region of the curve, the upper solid line is a guide to the eye. In the inset is shown the excess peak-height (data-points minus the Debye curve in each case) for both the $x = 0.25$ and $x = 0.21$ samples plotted vs. reduced temperature. They exhibit scaling behavior, though the $x = 0.21$ sample does not become fully delocalized even at low temperature so does not follow the $x = 0.l25$ curve all the way.

for the distorted polaronic phase may be the polaronic charge ordered phase observed at low-temperature in the $x = 0.5$ sample.[27] These results are interesting in light of a possible percolation model for the transport because,

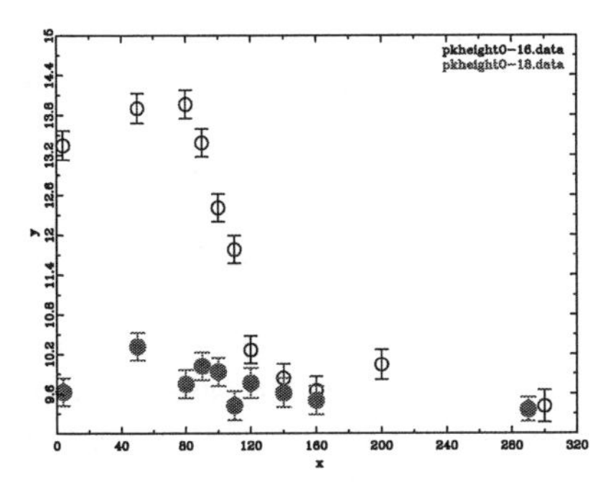

Figure 3. Plot of PDF peak-height of the $r = 0.275$ O-O peak (arbitrary units) vs. temperature for two samples of $La_{0.525}Pr_{0.175}Ca_{0.3}MnO_3$ enriched with ^{16}O (open circles) and ^{18}O (filled circles) respectively. In the ^{18}O the transition to a FM ground-state is suppressed and the sample remains an AF insulator to low temperature. This clearly shows that the excess PDF peak-height is coming from the appearance of undistorted octahedra in the metallic phase.

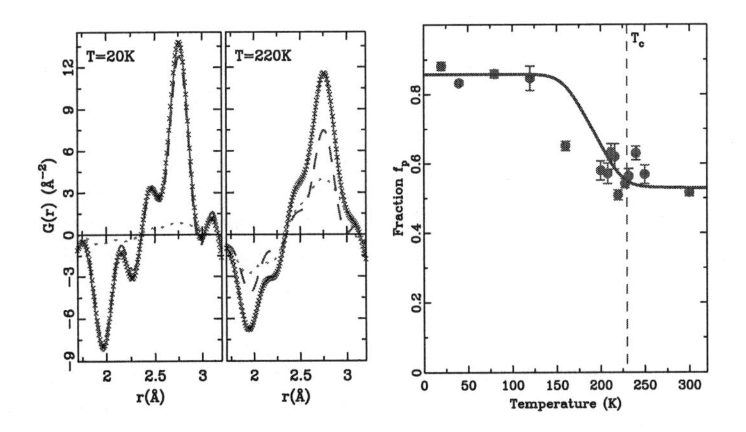

Figure 4. First two panels: low-r region of $La_{1-x}Ca_xMnO_3$ $x = 0.25$. Shown are data (symbols) best-fit of two-phase models (solid line) distorted phase (dotted line) and undistorted phase (dashed line). At low temperature (20K) the sample is predominantly undistorted, at 220K there is a significant proportion of both distorted and undistorted. Third panel: the resulting refined phase fractions as a function of temperature.

(a) If the sample is 50% metallic above T_{mi}, this already exceeds the geometric percolation threshold for the 3-D cubic lattice (though not the 2-D case) suggesting non-random percolation and (b) the sharp increase in the number of undistorted octahedra (associated with the metallic state) at T_{mi} would suggest a strong feedback mechanism. The metallic phase increases slowly with decreasing temperature until it percolates whence it begins to increase rapidly (e.g., see Fig. 2). Again, this suggests some kind of correlated percolation if the percolation picture is right at all. The origin of this behavior is not completely clear, but clearly the electronic properties depend on structures on multiple scales.

A similar competition between localized, polaronic and delocalized metallic behavior, is occurring in the $x = 0.5$ sample. Yet another length-scale comes into play in this case. In heavily doped materials the polarons prefer to order to minimize strain; so called charge order. In $La_{1-x}Ca_xMnO_3$ at $x = 0.5$ this occurs in zigzag stripes that results in CE magnetic order[13] so we will refer to it as CE charge order. The ground-state at $x = 0.5$ is antiferromagnetic insulating with CE charge order. However, this sample is very close to the FM phase at slightly lower doping and, indeed, on cooling the sample goes first ferromagnetic before becoming charge ordered at lower temperature. We have plotted the PDF peak height through these transitions as shown in Fig. 5.[28] On becoming ferromagnetic the $r = 2.75$ Å peak that showed the scaling behavior (Fig. 2) again sharpens on going through the ferromagnetic transition, following the scaling law which is plotted as a dashed line in Fig. 5; the sample is heading towards its ferromagnetic ground-state and gradually transforming into undistorted, delocalized metal. Nonetheless, much of the sample still remains polaronic. At $T = 180$ K this polaronic part charge-orders, apparent from the appearance of CO superlattice peaks at this temperature. Below this temperature, instead of the PDF peak-height curve following the scaling law curve towards complete metallicity it drops back to the average line (Fig. 5) and the part of the sample already transformed to metallic is restored to a polaronic state. The energy balance between these two states is very delicate apparently and the lowering in energy of the polaronic state due to it being permitted to order commensurately with the lattice is enough to stabilize it with respect to the FM phase. The length-scale for the charge order is roughly the stripe-separation which is around 10 Å. This is larger than the ~ 3 Å size of the polarons themselves and the nanometer length-scale (imprecisely determined) of the nano-phase separation. This picture is also consistent with the observation of short-range ordered charge-ordered clusters in the polaronic phase above T_{mi} at dopings below $x = 0.5$.[29]

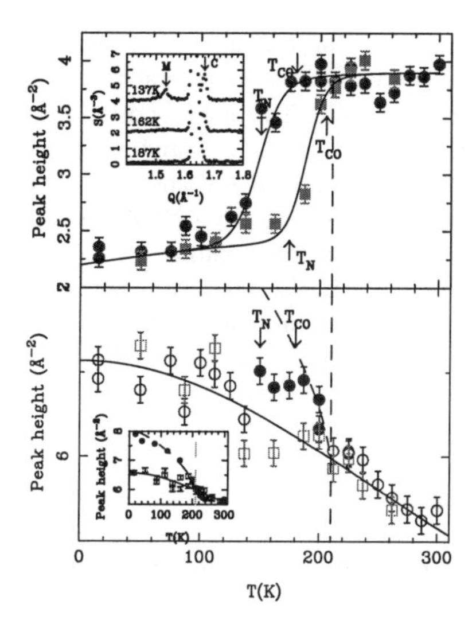

Figure 5. 50% PDF peakheight plots from $La_{1-x}Ca_xMnO_3$ $x = 0.5$ as a function of temperature. The top panel shows a high-r peak that is sensitive to the charge-ordering transition (indicated by arrows for cooling and warming) but not the ferromagnetic transition (shown by the dashed line). The lower curve shows the O-O peak at $r = 0.275$ Å that responds to the FM transition by sharpening, following the scaling curve discussed above and shown as a dashed line. However, when the portion of the sample that is still polaronic charge orders, the delocalized portion of the sample is quickly swallowed up the peak-height returns to the smooth curve of the distorted phase. This is a failed MI transition.

3. Cuprates

3.1. *Stripes and structural distortions*

Our investigations in the cuprates have focussed principally on the $La_{2-x}Sr_xCuO_4$ system. Here the PDF has been used to look for evidence of localized charges in direct analogy with the case of the manganites. The polaronic distortions in this case are thought to be much smaller, being somewhere between ~ 0.02 Å[30] and ~ 0.004 Å[31] making them 10-50 times smaller than in the manganites (though note that XAFS data have been interpreted as due to a larger distortion on a minority of sites.[32]) The small size of polaronic distortions in this system makes them difficult to study using structural probes and results should be considered only semi-quantitative. The size of the polaronic distortion predicted by Bozin *et al.*[30]

is based not directly on PDF measurements but taken from the magnitude of bond-shortening observed in the average structure due to doping and determined crystallographically. This is beyond the resolution of the PDF (and XAFS) which is around $r = 0.12$ Å and set by physical limits; i.e., the quantum zero-point motion of the atoms. Differences in bond-length shorter than this cannot be seen directly in PDF and XAFS data but only inferred by multi-peak fits to a single feature.

Intermediate range data is also present in powder diffraction derived PDFs. This contains information about such things as CuO_6 octahedral tilts. The presence of polarons or stripes has implications on the octahedral tilts and this can give complementary information about the presence or absence of such disorder. Below we briefly describe the PDF evidence for local short-range ordered stripes, though this has been extensively described in a series of publications.[30,33]

3.2. *Structural evidence for short-range ordered stripes*

First we develop an argument about the structural consequences of charge stripes, then we look in the PDF data for supporting evidence. The first observation is that doping charge into the planar Cu-O shortens them. This is a universally observed experimental fact in the cuprates and is easily understood since the planar bonds are σ^* antibonding bands and doping positive charge (holes) therein stabilizes the covalent bond and shortens it. Based on data of Radaelli *et al.*[34] the (average) bond shortening in $La_{2-x}Sr_xCuO_4$ is $\frac{dr_{Cu}}{dx} \sim -0.1$ Å/doped charge/copper, thus doping $x = 0.2$ leads to a bond shortening of ~ 0.02 Å. Now we assume the presence of stripes. This means 1-D stripey objects with increased doped charge on them separated by stripes with less doped charge. It immediately becomes obvious that the "charged stripes" must have shorter bonds than the "uncharged stripes" though they are topologically connected by covalent bonds. This has experimental and theoretical implications. The theoretical implications are that the existence of charge stripes introduces a misfit strain that tends to break the stripes up. This is discussed below. The experimental implications are that evidence should exist in structural probes for a distribution of planar Cu-O bond lengths. When the stripes are long-range ordered this will appear in the form of superlattice peaks in neutron diffraction as already observed.[31] If the stripes exist locally but are not long-range ordered we have to look using a probe sensitive to the local structure such as PDF. As already mentioned this difference in bond length cannot be directly observed so it is necessary to search for a doping dependent broadening in the planar Cu-O bond length distribution. This was observed in

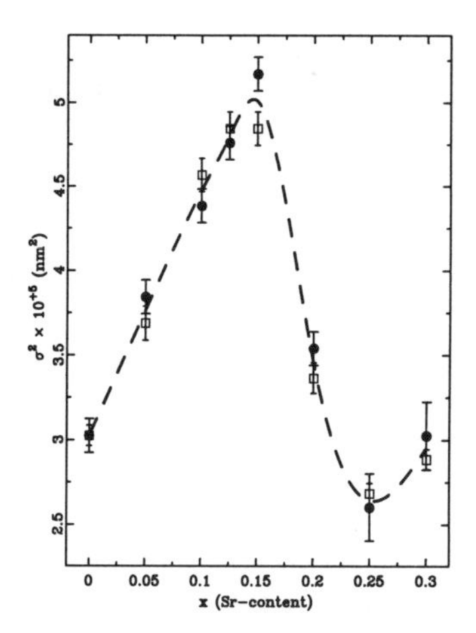

Figure 6. Peak width of the in-plane Cu-O PDF peak as a function of doping at 10K.

$La_{2-x}Sr_xCuO_4$,[30] peaking at $x = 0.15$ (rather than at $x = 0.5$ which would be expected if the charges are localized randomly on single copper sites). This is shown in Fig. 6(a). Above $x = 0.15$ the bond distribution sharpened again, presumably as the stripe order is replaced by a more homogeneous charge distribution. Temperature dependence of the bond distribution has also been studied. There is much scatter in the data, which testifies to the difficulty of the measurements due to the small polaronic distortion, but an anomalous broadening at low temperature (below $\sim$ 100 K) is evident[35] that suggests the appearance of charge stripes. This is consistent with XAFS data that shows a broadening in the planar Cu-O bond distribution at low temperature,[32] although the modelling of these data resulted in a different interpretation.[32]

There are other structural implications of the presence of stripes. The CuO_2 planes in $La_{2-x}Sr_xCuO_4$ are buckled by cooperative tilts of CuO_6 octahedra. You can shorten a Cu-O bond locally without straining it longitudinally by locally removing the octahedral tilt. The existence of short-bond charged stripes therefore also implies that there will be a distribution of CuO_6 octahedral tilt *angles.* The intermediate range region of the measured PDFs should be consistent with this if stripes are present. Indeed, a model that contained mixed tilt amplitudes reproduced the data very well[33,30] which, at least, shows that the observed PDFs are consistent with the presence of octahedral tilt disorder. Furthermore, this study[33,30] indicated that some tilt *directional* disorder was also present (i.e., a mixture of $\langle 110 \rangle$ and $\langle 100 \rangle$ symmetry tilts). This is expected from topological arguments.[36] The PDF data are therefore consistent with the presence of short-range ordered charge-stripes in the underdoped region of $La_{2-x}Sr_xCuO_4$ at low temperature.

3.3. *Multi-scale structure: stripe domains*

We new briefly describe the theoretical implications of the above discussion, and that is the appearance of an "interfacial" lattice misfit strain as a direct consequence of charge-stripe formation. The basic arguments that lead to this conclusion were laid out at the beginning of the last section and have been described in detail elsewhere.[37,36] When charge is doped into the CuO_2 plane the bonds shorten. The presence of stripes implies that regions of the plane coexist side-by-side that are heavily doped and lightly doped. These regions are topologically connected by covalent bonds. The charged stripes want to be shorter than the uncharged regions between and this is the origin of the misfit strain. As the stripe gets longer the strain increases. An infinite stripe would have an infinite strain energy, so at some characteristic length that is a balance between the stripe formation tendency (presumed to come from the electronic system and magnetism) and the strain energy and the stripes break up. A simple model that captures this physics is described in Refs. [37,36]. A lattice gas with attractive near neighbor (J_{nn}) and repulsive next nearest neighbor (J_{nnn}) interactions for doped sites ensures stripe formation. The strain terms come about due to a misfit of the desired bond-length of the doped site and the constraint imposed by the average periodic potential that it sits in, determined by the rest of the crystal.

The critical breakup length for the stripes is given by $L_c = N_c a$, where the critical number of copper sites, N_c, per strained stripe scales like

$$N_c \sim \left(\frac{J_{nn}}{k_{nn}}\right)\left(\frac{1}{a - l_0}\right)^2 . \tag{2}$$

Here, k_{nn} is the harmonic spring constant between nearest neighbor doped sites, a is the average separation of copper ions dictated by the average structure and l_0 is the bond length of the shorter, doped, Cu-O bonds in the stripe.

This inherent tendency towards stripe breakup due to lattice strain will give rise to microstructures with domains of broken stripes. A number of intuitive possibilities are shown in Fig. 7. The first relieves strain by matching up short charged stripes with longer uncharged stripes. This has significantly lower energy than the system without the broken stripes.

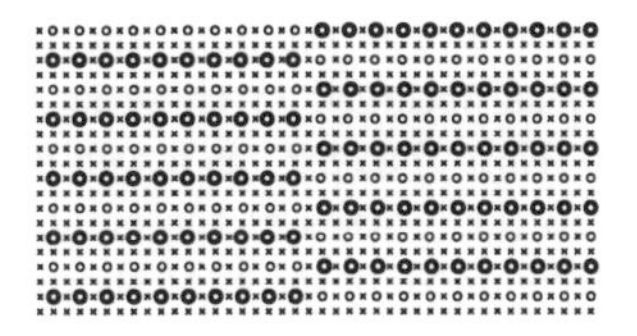

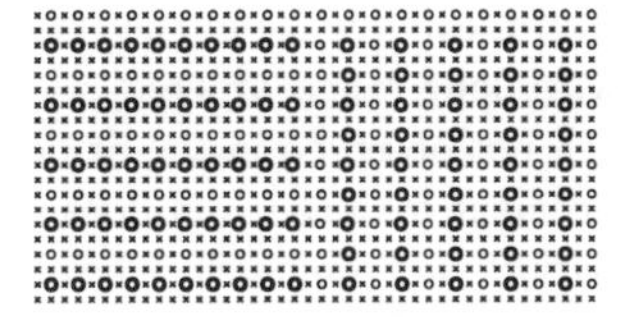

Figure 7. Stripe nanostructures in cuprates. (a) Interleaved stripes (b) weave microstructure. One unit cell of each microstructure is shown. Concentric circles indicate doped sites, while sites indicated by crosses are excluded from the strain relaxation. For illustration purposes the natural length of bonds between doped sites are 10% shorter than bonds between undoped sites.

It has similar energy to that shown in Fig. 7(a) that has the additional advantage that it is more isotropic and allows better strain relaxation in two dimensions. It should be noted that details of the magnetism, not included in this model, will impact the preferred microstructure because of spin frustration in the interfacial regions between domains. However, the general picture that strain breaks up stripes into a microstructure with a characteristic length-scale that is given approximately by Eq. 2 is robust.

This general picture may explain some of the phenomenology of the cuprates. If we assume that static, long-range ordered, stripes compete with superconductivity but short-range ordered dynamic stripes do not, or that fluctuating stripes even enhance superconductivity,[38] we can make the following observations. A longer length-scale for stripe breakup will result in more slowly fluctuating stripes, and a poorer superconductor, than a material with a shorter stripe breakup length-scale. A shorter length-scale appears in systems with more strain. At this point we haven't discussed how this interfacial strain can be relaxed; however, a number of possibilities exist. For example, in the $La_{2-x}Sr_xCuO_4$ system the CuO_2 planes are buckled. Part of the bond-shortening required when a copper is doped can be accommodated by locally driving away the tilt without straining the Cu-O covalent bond.[37] Thus, we might expect that in this system the misfit strain that leads to stripe breakup is less than that in a system with flat planes, for example. Longer stripes result, a larger microstructure and slower fluctuations. In the extreme case where the sample is co-doped with misfitting Nd, or doped with misfitting Ba, there may be sufficient structural compliance in the tilts to accommodate all the bond shortening allowing long-range ordered static stripes as observed in these systems, though not observed in the Sr doped case. Note that Sr^{2+} is a well-fitting replacement for La^{3+} and does not perturb (i.e., increase) the octahedral tilt background too much. The ability to accommodate the bond shortening without a resulting misfit strain we call "structural compliance".

There are other possible sources of structural compliance in the cuprates. For example, in the YBCO system, chain oxygen atoms have the opportunity to self organize so as to minimize the misfit strain in the cuprate planes. Chain-oxygen ordering has a well documented effect on T_c in these systems[39] though this was hitherto thought to be due to charge-transfer effects. Similar effects are seen due to interstitial oxygen ordering in $La_2CuO_{4+\delta}$ which again may be related to self organization to minimize the energy of the stripe microstructure.

A general observation is that as the CuO_2 planes get flatter (and therefore the tilting source of structural compliance disappears), T_c goes up. Focussing attention on the materials with a single CuO_2 layer it is observed that the highest T_c material is $HgBa_2CuO_6$. This has flat planes and a very simple structure with few possibilities for structural compliance. Likewise, the single layer thallium compound which is also a high-T_c material, has flat planes and just a bit of structural disorder in the out-of-plane layers[40] that could self-organize. Also a high-T_c material but with a little lower optimal T_c, is the bismuth material that relaxes a mismatch between the CuO_2 and intergrowth layers with an incommensurate structural modulation.[41] YBCO is a two-layer system with a moderate T_c (similar to the single layer mercury compound and the lowest of the two-layer bismuth, thallium or mercury materials) and it also has buckled CuO_2 planes.

These empirical observations are at least qualitatively explainable within the picture of strain induced stripe-breakup and microstructure due to misfit strain. They point to the importance of engineering structures on multiple different length-scales in order to optimize electronic properties in these materials. Here it seems the atomic scale is (as always) important, but also the stripe length-scale of ~ 10 Å and the length-scale of the stripe microstructure that can vary from short (maybe comparable to the stripe spacing) all the way to micron sized, with comparable change in material properties.

4. Conclusions

Here we have summarize investigations of the local atomic structure in the transition metal cuprates and manganites. Nano-scale electronic inhomogeneities appear to be widespread and can have structures on a number of different length-scales. These multi-scale structures have a profound effect on the electronic properties of these materials. In the case of the cuprates we have presented a model for lattice-strain induced stripe breakup. The lattice strain is an inevitable consequence of having a microscopically inhomogeneous charge distribution and has analogs in more systems which support variable doping and possible charge inhomogeneities.

Acknowledgments

None of this work would have been possible without the tireless efforts of past and present members of the Billinge group: Emil Božin, Matthias Gutmann, Thomas Proffen, Valeri Petkov, Peter Peterson, Il-Kyoung Jeong and Xiangyun Qiu. It also benefitted from financial support from NSF through grant DMR-0075149. The results presented made use of a number of x-ray and neutron facilities: IPNS at Argonne National Laboratory (DOE-BES contract number W-31-109-Eng-38), MLNSC, Los Alamos National Laboratory (DOE contract W-7405-ENG-36), APS, Argonne National Laboratory (DOE BES contract W-31-109-Eng-38).

References

1. E. K. H. Salje, Contemp. Phys. **41**, 79 (2000).
2. S. R. Shenoy, T. Lookman, A. Saxena, and A. R. Bishop, Phys. Rev. B **60**, R12537 (1999).
3. B. P. Stojković, Z. G. Yu, A. L. Chernyshev, A. R. Bishop, A. H. C. Neto, and N. Gr onbech-Jensen, Phys. Rev. B **62**, 4353 (2000).
4. K. O. Rasmussen, T. Lookman, A. Saxena, A. R. Bishop, R. C. Albers, and S. R. Shenoy, Phys. Rev. Lett. **87**, 055704 (2001).
5. A. H. Castro Neto, Phys. Rev. B **64**, 104509 (2001).
6. M. Uehara, S. Mori, C. H. Chen, and S.-W. Cheong, Nature **399**, 560 (1999).
7. S. H. Pan, J. P. O'Neal, R. L. Badzey, C. Chamon, H. Ding, J. R. Engelbrecht, Z. Wang, H. Eisaki, S. Uchida, A. K. Gupta, K.-W. Ng, E. W. Hudson, K. M. Lang, and J. C. Davis, Nature **413**, 282 (2001).
8. R. Prinz and D. Koningsberger, editors, *X-ray absorption: principles, applications techniques of EXAFS, SEXAFS and XANES*, J. Wiley and Sons, New York, 1988.
9. T. Egami and S. J. L. Billinge, *Underneath the Bragg Peaks: Structural analysis of complex materials*, Pergamon, Oxford, England, 2002.
10. T. Egami and S. J. L. Billinge, Prog. Mater. Sci. **38**, 359 (1994).
11. T. Egami and S. J. L. Billinge, in *Physical properties of high-temperature superconductors V*, edited by D. M. Ginsberg, page 265, Singapore, 1996, World–Scientific.
12. C. Zener, Phys. Rev. **82**, 403 (1951).

13. J. B. Goodenough, Phys. Rev. **100**, 564 (1955).
14. A. J. Millis, P. B. Littlewood, and B. I. Shraiman, Phys. Rev. Lett. **74**, 5144 (1995).
15. H. Röder, J. Zang, and A. R. Bishop, Phys. Rev. Lett. **76**, 1356 (1996).
16. S. J. L. Billinge, R. G. DiFrancesco, G. H. Kwei, J. J. Neumeier, and J. D. Thompson, Phys. Rev. Lett. **77**, 715 (1996).
17. S. J. L. Billinge, Th. Proffen, V. Petkov, J. Sarrao, and S. Kycia, Phys. Rev. B **62**, 1203 (2000).
18. C. H. Booth, F. Bridges, G. J. Snyder, and T. H. Geballe, Phys. Rev. B **54**, R15606 (1996).
19. D. Louca, T. Egami, E. L. Brosha, H. Röder, and A. R. Bishop, Phys. Rev. B **56**, R8475 (1997).
20. D. Louca and T. Egami, Phys. Rev. B **59**, 6193 (1999).
21. A. Moreo, A. Yunoki, and E. Dagotto, Science **283**, 2034 (1999).
22. M. Jaime, P. Lin, S. H. Chun, M. B. Salamon, P. Dorsey, and M. Rubinstein, Phys. Rev. B **60**, 1028 (1999).
23. M. Mayr, A. Moreo, J. A. Vergés, J. Arispe, A. Feiguin, and E. Dagotto, Phys. Rev. Lett. **86**, 135 (2001).
24. G. H. Kwei, A. C. Lawson, S. J. L. Billinge, and S.-W. Cheong, J. Phys. Chem. **97**, 2368 (1993).
25. D. Kim, B. Revaz, B. L. Zink, F. Hellman, J. J. Rhyne, and J. F. Mitchell, (2002), cond-mat/0210088.
26. P. G. Radaelli and D. Argyriou, private communication.
27. P. G. Radaelli, G. Iannone, M. Marezio, H. Y. Hwang, S.-W. Cheong, J. D. Jorgensen, and D. N. Argyriou, Phys. Rev. B **56**, 8265 (1997).
28. S. J. L. Billinge, R. G. DiFrancesco, M. F. Hundley, J. D. Thompson, and G. H. Kwei, Phys. Rev. Lett. (2000), Unpublished.
29. J. W. Lynn, C. P. Adams, Y. M. Mukovskii, A. A. Arsenov, and D. A. Shulyatev, J. Appl. Phys. **89**, 6846 (2001).
30. E. S. Božin, S. J. L. Billinge, H. Takagi, and G. H. Kwei, Phys. Rev. Lett. **84**, 5856 (2000).
31. J. M. Tranquada, J. D. Axe, N. Ichikawa, Y. Nakamura, S. Uchida, and B. Nachumi, Phys. Rev. B **54**, 7489 (1996).
32. A. Bianconi, N. L. Saini, A. Lanzara, M. Missori, T. Rossetti, H. Oyanagi, H. Yamaguchi, K. Oka, and T. Ito, Phys. Rev. Lett. **76**, 3412 (1996).
33. E. S. Božin, S. J. L. Billinge, G. H. Kwei, and H. Takagi, Phys. Rev. B **59**, 4445 (1999).
34. P. G. Radaelli, D. G. Hinks, A. W. Mitchell, B. A. Hunter, J. L. Wagner, B. Dabrowski, K. G. Vandervoort, H. K. Viswanathan, and J. D. Jorgensen, Phys. Rev. B **49**, 4163 (1994).
35. M. Gutmann, E. S. Božin, and S. J. L. Billinge.
36. S. J. L. Billinge and P. M. Duxbury, Int. J. Mod. Phys. B (2002).
37. S. J. L. Billinge and P. M. Duxbury, **66**, 064529 (2002).
38. V. J. Emery, S. A. Kivelson, and J. M. Tranquada, Proc. Natl. Acad. Sci. USA **96**, 8814 (1999).
39. H. Shaked, J. D. Jorgensen, B. A. Hunter, R. L. Hitterman, A. P. Paulikas, and B. W. Veal, Phys. Rev. B **51**, 547 (1995).
40. B. H. Toby, T. Egami, J. D. Jorgensen, and M. A. Subramanian, Phys. Rev. Lett. **64**, 2414 (1990).
41. A. Yamamoto, M. Onoda, E. TakayamaMuromachi, F. Izumi, T. Ishigaki, and H. Asano, Phys. Rev. B **42**, 4228 (1990).

The Chain Layer of YBCO: a close-up view with STM

Alex de Lozanne

Department of Physics, University of Texas
Austin, TX 78712-1081.

A brief review is given of results obtained by scanning tunneling microscopy and spectroscopy on the chain layer of YBCO, with general comments on this technique. An informal personal view on the high Tc community is also given.

An ancient story

The experimentalists studying the high tc materials remind the author of the ancient story of the six blind men who were asked to describe a large object.[1] Their descriptions were very different, depending on the detail on which each man focused: a wall, a spear, a snake, a tree, a fan, and a rope.

The moral of this parable is beautifully expressed in poetic fashion [1]

So oft in theologic wars,
The disputants, I ween,
Rail on in utter ignorance
Of what each other mean,
And prate about an Elephant
Not one of them has seen!

Many of our colleagues take data in reciprocal space with techniques such as neutron or x-ray diffraction. The raw data is not intuitive to interpret, so not even experts would look at Fig 1a and recognize it as an elephant (in Fourier space!). On the opposite end of the spectrum, those of us who work in "real space" have an advantage because our pictures are eye-catching and intuitive, yet we often look at our objects so closely that we do not get the whole picture either. Fig. 1b is also an elephant, now in real space, but at such a high magnification that nobody would recognize it. Clearly, we need both real space and reciprocal space probes at all relevant scales to understand the object we study, in this example depicted by Fig. 1c. Perhaps more challenging than performing our difficult experiments is to understand the other blind men and women working on these compounds. That is why workshops such as this one are important.

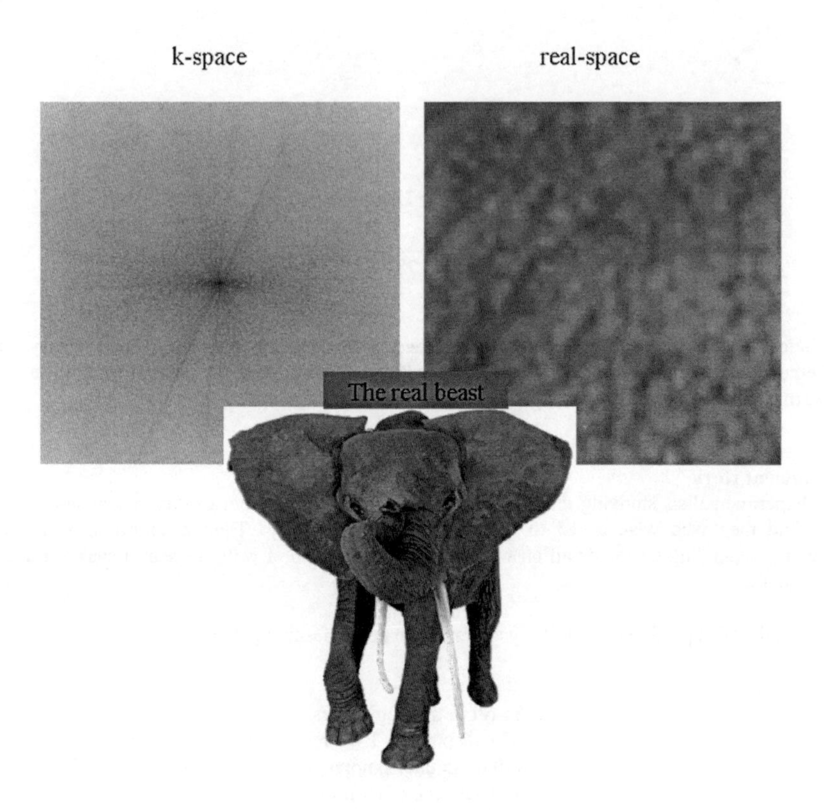

Figure 1
An elephant in reciprocal space, real space, and reality.

A tame animal and a ferocious beast, all in one.

On one hand, the high Tc materials are so easy to tame that high school students (and younger) can make a decent superconductor with Tc above 90 K. These structures are clearly very robust against impurities and defects, more so than niobium and its compounds. On the other hand, those of us trying to understand these materials have found them to be far from tamable, to the point that only a few labs in the world can grow high quality crystals, and not many have learned to obtain meaningful data from them.

This situation is particularly true for measurements with a scanning tunneling microscope (STM), which probes the local density of states at the surface of a sample. The first STM measurements of a superconductor were published in 1984-85.[2] Traditionally surfaces are prepared by ion bombardment to remove contaminants, followed by in-situ annealing to remove the damage caused by the ions. This is not possible with the high Tc materials because their unit cells are very large, containing at least four different atomic species. The scrambling of the unit cells on the surface cannot be repaired by in-situ (vacuum) annealing because these materials are extremely sensitive to oxygen stoichiometry. The oxygen loss from the surface would make the material under study quite different from the bulk, in an uncontrolled manner. Even if this could be prevented, the other elements are likely to have differing sputtering yields, so that the surface composition, other than oxygen, would be altered.

Fortunately some of the high tc materials can be cleaved to produce surfaces useful for STM and other techniques, and some surfaces can be exposed to air for a brief time, although this is not ideal. The chain layer of YBCO, however, must be cleaved and kept below 25K, otherwise oxygen desorbs and destroys the surface chains. Once we figured out how to cleave the surface of YBCO we published several papers on the subject.[2 - 12]

In the case of tunneling spectroscopy of superconductors the high T_C materials present yet another challenge: the superconducting properties are probed within one coherence length from the surface.[13] The challenge is that the coherence length is very short in these materials, so that the superconducting properties of the material must be good to within 1 nm or less from the surface.

For all these reasons from one must be very cautious to accept spectroscopic data taken in a point contact mode. The safest set of data is one where atomic resolution images are shown and the spectroscopic data is taken under the same conditions, namely with the same average tunneling conductance.

Modulations in the chains of YBCO.

Figure 2 shows the first and the last images of the YBCO chain layer that we have published. While the resolution has increased dramatically due to the STM design by S.H. Pan, the main features are present in both. First, The chains are very evident, running approximately along the diagonal in both images. Large dark depressions are believed to be oxygen vacancies. More of these depressions are seen in Fig. 2a than in 2b, because the sample is at a higher temperature in 2a. Along each chain, there are also bright and dark regions with a periodicity 3-4 times longer than the unit cell. We first realized that these modulations are due to electronic effects when we reversed the direction of tunneling.[3] Our latest results show that these modulations have a wavelength that depends on energy, so that they are likely Friedel oscillations produced by electron scattering from defects such as oxygen vacancies. A recent surprise was the discovery of sharp resonances or peaks in the tunneling spectra (Local density of states) at energies approximately inside the gap. The current understanding of these resonances is that they are due to scattering of quasiparticles from defects that have some magnetic character (such as a missing oxygen atom in an otherwise filled chain of YBCO).

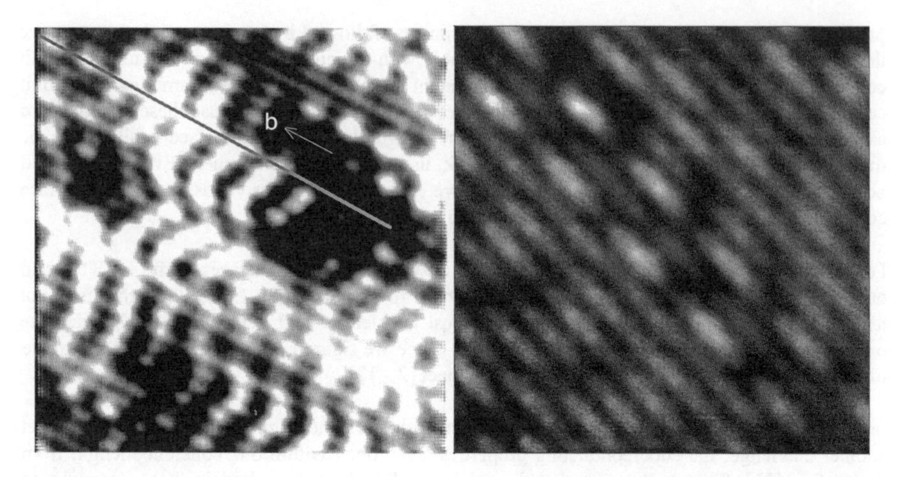

Figure 2
STM images of YBCO chain layer published in 1992 (left)[2] and 2002 (right)[12]. The respective sizes are 7.5 and 5 nm on a side.

An unfinished story

While we have learned many interesting and important details about this surface, much more remains to be explored. The effects of oxygen doping, especially in cases where the oxygen is ordered, are a natural next step. Other compounds with chains may also shed light on the general problem of having a one-dimensional conductor coupled to a superconductor. Of course, there is still the question about *The Mechanism* for high T_c superconductivity. While there has been unquestionable progress in our understanding of this question, our colleagues on the theory side sometimes seem to be in worse shape than our six blind men.

Acknowledgements

The results reviewed here represent the work of many people in the author's laboratory and in the laboratories of collaborators at UT Austin (J. Markert's group) and UC Berkeley (J.C. Davis' group). This work is supported by the NSF.

References

1 A poetic version of this ancient parable was written by John Godfrey Saxe (1816-1887). It can be read at:

http://elephant.elehost.com/About_Elephants/Stories/Parables/Blind_Men___Elephant/blind_men___elephant.html

2 H. L. Edwards, J. T. Markert and A. L. de Lozanne. Phys. Rev. Lett. **69**, 2967-2970 (1992).

3 H. L. Edwards, J. T. Markert, and A. L. de Lozanne, J. Vac. Sci. Tech. B **12**, 1886 (1994).

4 H. L. Edwards, A. L. Barr, J. T. Markert, and A. L. de Lozanne, *Phys. Rev. Lett.* **73**, 1154 (1994).

5 H. L. Edwards, D. J. Derro, A. L. Barr, J. T. Markert, and A. L. de Lozanne, *Phys. Rev. Lett.* **75**, 1387 (1995).

6 H. L. Edwards, D. J. Derro, A. L. Barr, J. T. Markert, and Alex de Lozanne, *J. Phys. Chem. Solids* **56**, 1803-1804 (1995).

7 H. L. Edwards, D. J. Derro, A. L. Barr, J. T. Markert, and A. L. de Lozanne, J. Vac. Sci. Technol. B **14**(2) 1217-1220 (1996)

8 A.L. de Lozanne, S. H. Pan, H. L. Edwards, D. J. Derro, A. L. Barr, and J. T. Markert, SPIE Vol. 2696, 347 (1996).

9 D. J. Derro, T. Koyano, Hal Edwards, A. Barr, J. T. Markert, and A. L. de Lozanne, in *Proc. of the 10th Anniversary HTS Workshop on Physics, Materials and Applications*, ed. by B. Batlogg, C.W. Chu, W.K. Chu, D.U. Gubser and K.A. Müller (World Scientific, Singapore, 1996) pg.433.

10 Alex de Lozanne, Superconductor Science and Technology **12**, R43-R56 (1999).

11 D. J. Derro, E. W. Hudson, K. M. Lang, S. H. Pan, J. C. Davis, K. Mochizuki, J. T. Markert, and A. L. de Lozanne, Physica C **341**, 425-428 (2000).

12 D.J. Derro, E.W. Hudson, K. M. Lang, S. H. Pan, J.C. Davis, J.T. Markert & A.L. de Lozanne, Phys. Rev. Let. **88**, 097002 (2002).

13 "Principles of Tunneling Spectroscopy", E.L. Wolf (Oxford Univ. Press, NY, 1985)

STRIPES AND CHARGE TRANSPORT PROPERTIES OF HIGH-T_C CUPRATES

YOICHI ANDO
Central Research Institute of Electric Power Industry,
Komae, Tokyo 201-8511, Japan
E-mail: ando@criepi.denken.or.jp

Unusual features in the in-plane charge transport in lightly hole-doped $La_{2-x}Sr_xCuO_4$ single crystals are described. Notably, both the in-plane resistivity and the Hall coefficient show a metallic behavior at moderate temperatures even in the long-range-ordered antiferromagnetic phase, which obviously violates the Mott-Ioffe-Regel criterion for the metallic transport and can hardly be understood without employing the role of charge stripes. Moreover, the mobility of holes in this "metallic" antiferromagnetic state is found to be virtually the same as that in optimally-doped crystals, which strongly suggests that the stripes govern the charge transport in a surprisingly wide doping range up to optimum doping.

1. Introduction

In high-T_c cuprates such as $La_{2-x}Sr_xCuO_4$ (LSCO), the antiferromagnetic (AF) state gives way to high-T_c superconductivity when a sufficient number of holes are doped into the CuO_2 planes. The AF state of cuprates is therefore a natural starting point to establish the picture of high-T_c superconductors, but nevertheless their transport properties have not drawn sufficient attention. It has been generally believed that the hole motion inevitably frustrates the antiferromagnetic bonds and thus the doped holes must be strongly localized until the long-range AF order is destroyed. Indeed, the variable-range-hopping conductivity has been mostly observed in the AF state of cuprates,[1,2] which is naturally expected for the localized holes. As a result, researchers have been discouraged by the apparent simplicity of this so-called "antiferromagnetic insulator" regime.

However, recent measurements in clean, lightly-doped $YBa_2Cu_3O_y$ (YBCO) crystals have demonstrated[3,4] that the charge transport in the AF state is full of surprise: the temperature dependence of the in-plane resistivity ρ_{ab} remains to be metallic (ρ_{ab} decreases with decreasing temperature) across the Néel temperature T_N, anomalous features in the magnetoresistance imply that holes form stripes instead of being homogeneously

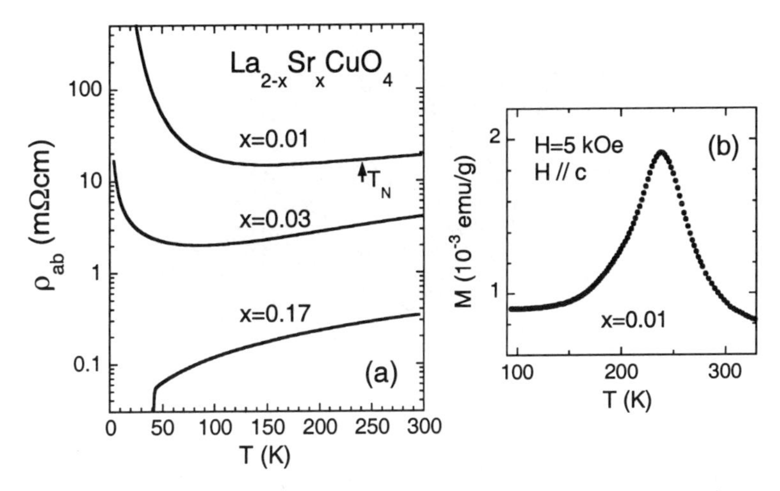

Figure 1. (a) Temperature dependences of ρ_{ab} of lightly-doped ($x = 0.01$ and 0.03) and optimally-doped ($x = 0.17$) $La_{2-x}Sr_xCuO_4$ single crystals. (b) Magnetization of a large $La_{1.99}Sr_{0.01}CuO_4$ single crystal from which the samples for ρ_{ab} measurements were cut; the peak in $M(T)$ corresponds to the Néel temperature.

distributed, and along the c-axis the charge confinement characteristics are significantly affected by the Néel ordering. Motivated by these results[3,4] on YBCO that we obtained in 1999, we have revisited the charge transport in clean single crystals of LSCO, where studying the lightly-doped regime is much more straightforward than in other cuprates; the hole doping p in the CuO_2 planes is equal to x, the Sr content, and T_N can be readily determined by susceptibility measurements.[5]

Here we show that, contrary to the common belief, the doped holes in clean single-crystalline cuprates are surprisingly mobile in a wide range of temperatures even in the long-range-ordered AF phase. This is possible when the electron system self-organizes into hole-rich stripes and hole-poor AF regions to facilitate the motion of charges. We further show that the hole mobility at moderate temperatures remains virtually unchanged throughout a wide doping range from the lightly-doped AF regime (hole doping of 1%) to the optimally-doped regime (hole doping of 17%) where the superconducting transition temperature is maximal. This strongly suggests that the hole motion is governed by the stripes all the way up to optimum doping, and thus the high-temperature superconductivity appears to be a property associated with the stripes.

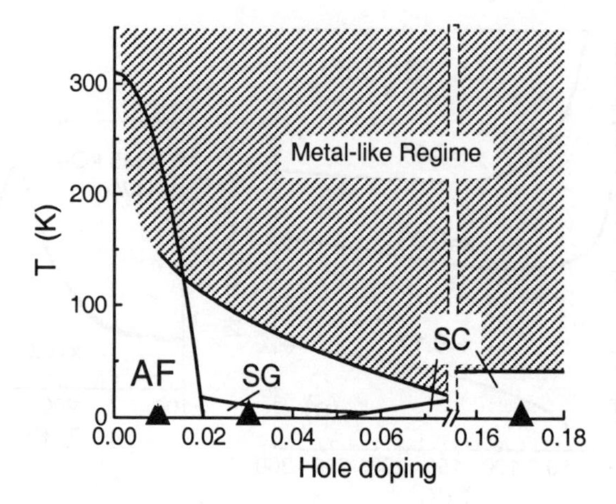

Figure 2. The antiferromagnetic (AF), spin-glass (SG) and superconducting (SC) regions on the phase diagram of LSCO; representative doping levels chosen for this article are indicated by triangles. The hatched region illustrates where ρ_{ab} shows the metal-like behavior ($d\rho_{ab}/dT > 0$).

2. Experimental

The clean single crystals of LSCO are grown by the traveling-solvent floating-zone (TSFZ) technique[6] and are carefully annealed to remove excess oxygen, which ensures that the hole doping is exactly equal to x. The in-plane resistivity ρ_{ab} and the Hall coefficient R_H are measured using a standard ac six-probe method. The Hall effect measurements are done by sweeping the magnetic field to ± 14 T at fixed temperatures stabilized within ~ 1 mK accuracy.[3] The Hall coefficients are always determined by fitting the H-linear Hall voltage in the range of ± 14 T, which is obtained after subtracting the magnetic-field-symmetrical magnetoresistance component caused by small misalignment of the voltage contacts.

3. Results

Figure 1(a) shows the temperature dependences of ρ_{ab} for LSCO crystals which represent three doping regimes[1,7] on the phase diagram (Fig. 2): antiferromagnetic [the sample with $x = 0.01$ has $T_N \simeq 240$ K according to the magnetization data shown in Fig. 1(b)], spin glass ($x = 0.03$), and optimally-doped superconductor ($x = 0.17$). (More complete data sets can be found in our recent papers.[6,8]) One may notice that, while the magnitude of the resistivity significantly increases with decreasing doping, the

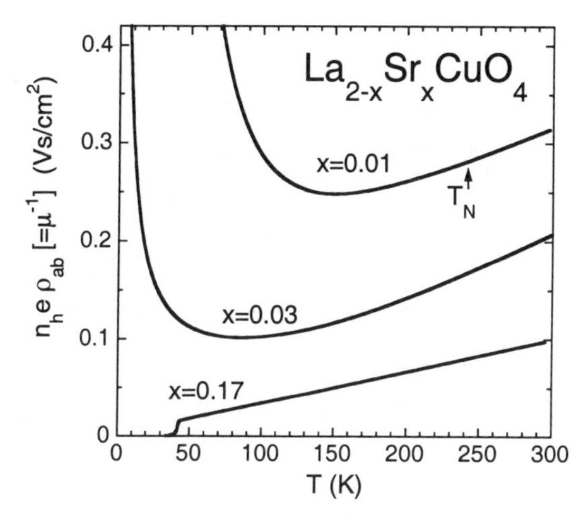

Figure 3. Temperature dependences of the normalized resistivity $n_h e \rho_{ab}$ of LSCO crystals, where $n_h = 2x/V$ is the nominal hole density. Note that $n_h e \rho_{ab}$ is essentially an inverse mobility μ^{-1} of doped holes.

temperature dependence at $T > 150$ K does not change much; in particular, in the sample with $x = 0.01$, ρ_{ab} keeps its metallic behavior well below T_N. This observation in the lightly-doped LSCO crystal clearly invalidates the long-standing notion that the metal-like behavior of $\rho_{ab}(T)$ in cuprates may appear only as soon as the long-range AF order is destroyed.

To examine whether the hole mobility actually depends on the magnetic state as crucially as has been expected, in Fig. 3 we normalize ρ_{ab} by the nominal hole concentration n_h, which is given by $2x/V$ [unit cell V ($\simeq 3.8 \times 3.8 \times 13.2$ Å^3) contains two CuO_2 planes]. The product $n_h e \rho_{ab}$ would mean just inverse hole mobility μ^{-1} if we assume the number of mobile holes to be always given by x. Apparently, the slope and magnitude of $n_h e \rho_{ab}$ at moderate temperatures are very similar, suggesting that the transport is governed by essentially the same mechanism for all three doping regimes; in particular, the magnitudes of the hole mobility at room temperature differ by only a factor of three between $x = 0.01$ and 0.17, demonstrating that the hole mobility remains virtually unchanged in a surprisingly wide range of doping. We note that the magnitude of the hole mobility in LSCO (order of 10 cm^2/Vs at 300 K) is almost the same as that in YBCO;[8] this suggests that the hole mobility in the CuO_2 planes is essentially universal among the cuprates. Interestingly, typical metals (such as iron or lead) show similar values of carrier mobility, $(ne\rho)^{-1}$, at room temperature.[8]

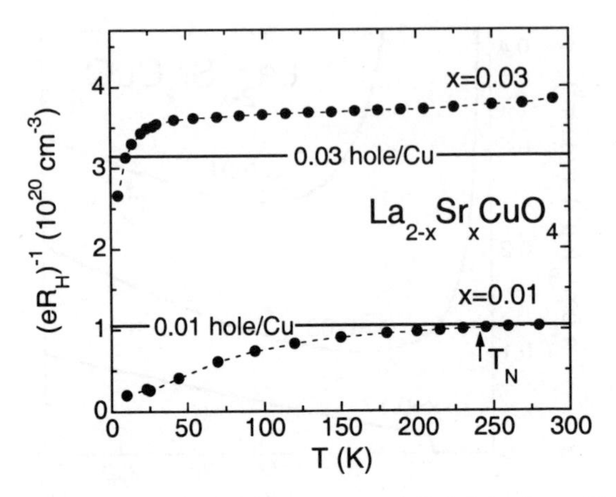

Figure 4. The apparent hole density of carriers $n = (eR_H)^{-1}$ for the two lightly-doped LSCO crystals; solid lines indicate the nominal value n_h.

The region that is characterized by the metallic transport behavior ($d\rho_{ab}/dT > 0$) is depicted in the phase diagram (Fig. 2); evidently, it extends widely in the phase diagram and essentially ignores the changes in the magnetic properties. It is worth noting that the normal-state resistivity in superconducting LSCO was studied[9,10] by suppressing superconductivity with 60-T magnetic fields and an increase in ρ_{ab} at low temperature was observed up to optimum doping; thus, the high mobility of holes at moderate temperature and localization at low temperature appear to be essentially unchanged in the normal state in the whole underdoped region, all the way from $x = 0.01$ to 0.15.

Another evidence for unexpected metallic charge transport in the AF cuprates can been found in the Hall coefficient R_H. The apparent hole density $n = (eR_H)^{-1}$ obtained for the LSCO samples with $x = 0.01$ and 0.03 (Fig. 4) is essentially temperature independent in the temperature range where the metallic behavior of $\rho_{ab}(T)$ is observed, which is exactly the behavior that ordinary metals show. Moreover, n agrees well with the nominal hole concentration $n_h = 2x/V$ at $x = 0.01$, which means that all the doped holes are moving and contributing to the Hall effect even in the long-range-ordered AF state down to not-so-low temperatures until disorder causes the holes to localize. For higher doping, the ratio n/n_h exceeds unity and reaches a value of $\sim$3 at optimum doping.

4. Discussions

4.1. *Unusual Metallic Transport*

It is useful to note that the absolute value of ρ_{ab} for $x = 0.01$ is as large as 19 mΩcm at 300 K. If we calculate the $k_F l$ value (k_F is the Fermi wave number and l is the mean free path) using the formula $hc_0/\rho_{ab}e^2$ (c_0 is the interlayer distance), which implicitly assumes a *uniform* 2D electron system and the Luttinger's theorem, the $k_F l$ value for $x = 0.01$ would be only 0.1; this strongly violates the Mott-Ioffe-Regel limit for metallic transport, and thus the conventional wisdom says that the band-like metallic transport is impossible for $x = 0.01$. In other words, the metallic transport in the slightly hole-doped LSCO is a strong manifestation of the "bad metal" behavior.[11]

Very recent angle-resolved photoemission spectroscopy (ARPES) measurements of lightly-doped LSCO crystals have found[12] that "Fermi arcs" develop at the zone-diagonal directions in the **k**-space, on which metallic quasiparticles are observed. These Fermi arcs are different from the small Hall pockets and apparently violate the Luttinger's theorem, because the Fermi surface is partially destroyed and thus the enclosed *area* is not well-defined. Therefore, at least phenomenologically, such violation of the Luttinger's theorem by the Fermi arcs allows the system to have a small effective carrier number and a "large" k_F value at the same time, which enables the metallic transport to be realized in the lightly hole-doped regime. (Thus, the $k_F l$ value estimated under the assumption of a uniform 2D system is obviously erroneous.)

4.2. *Difficulty of Metallic Transport in the Antiferromagnetic State*

How can such an unusual metallic transport and the relatively high mobility of doped holes be possible in the long-range-ordered AF phase? It has been known for a long time that a single hole doped into a two-dimensional square antiferromagnet should have a very low mobility because of the large magnetic energy cost of the spin bonds broken by the hole motion, although quantum effects allow the hole to propagate.[13,14] Despite this common knowledge, our resistivity and the Hall coefficient data demonstrate that the doped holes in the AF state can have the mobility nearly as high as that at optimum doping, which means that the holes manage to move without paying the penalty for frustrating AF bonds. This striking contrariety is not restricted to the simple one-band model implicitly hypothesized in the above argument. Whatever the transport mechanism is, the doped holes should have an extremely strong coupling to the AF background; otherwise such a small amount of holes as 2% would not be able to destroy the AF state.[1] At the same time, this strong coupling tends to localize the holes arbitrarily distributed in the AF background, since the spin distortion created by a hole in the rigid Néel state destroys the translational symmetry. Therefore, the unusually metallic charge transport in the AF phase requires a novel mechanism to be realized in the lightly-doped cuprates.

4.3. *Role of Stripes*

To the best of our knowledge, the only possibility for the metal-like conductivity to survive under the strong coupling of holes with the magnetic order is when the holes and spins form a superstructure which restores the translational symmetry. A well-known example is the striped structure,[14,15,16] where the energy cost for the distortion of the spin lattice is paid upon the stripe formation and then the holes can propagate along the stripes without losing their kinetic energy. In fact, the striped structure has been already established[17] for $La_{2-x-y}Nd_ySr_xCuO_4$, and there is now growing evidence for the existence of stripes in other hole-doped cuprates,[3,18,19,20,21] the case being particularly strong for LSCO and YBCO in the lightly-doped region. Moreover, the mesoscopic phase segregation into the metallic paths (charge stripes) and the insulating domains (AF regions) offers a natural explanation about why the apparent $k_F l$ value can be so small in the regime where metallic transport is observed.[8] Existence of such charged magnetic-domain boundaries are actually indicated by our recent in-plane anisotropy measurements of the magnetic susceptibility of lightly-doped LSCO.[22]

One might wonder about the nature of the Hall effect when the conductivity occurs through the quasi-one-dimensional (1D) stripes. Indeed, it was shown that the Hall effect tends to disappear in $La_{1.4-x}Nd_{0.6}Sr_xCuO_4$ (LNSCO) upon the transition into the static stripe phase.[23] Against our intuition, however, the quasi-1D motion itself does *not* necessarily drive the Hall coefficient to zero. The quasi-1D confinement dramatically suppresses the transverse (Hall) *current* induced by the magnetic field, but the same large transverse resistivity restores the finite Hall *voltage*, because $R_H \sim \sigma_{xy}/\sigma_{yy}\sigma_{xx}$. For the same reason, for instance, the well-known charge confinement in the CuO_2 planes in cuprates does not prevent generation of the Hall voltage along the c-axis ($H \parallel ab$).[24] Therefore, the Hall-effect anomaly in LNSCO must be caused by some more elaborate mechanism rather than simply due to the quasi-1D nature of the transport. One possibility is that the anomaly in LNSCO is due to the peculiar arrangement of stripes which alter their direction from one CuO_2 plane to another and thereby keeping σ_{yy} from vanishing; on the other hand, the unidirectional stripes[25] in pure LSCO would naturally keep the Hall coefficient unchanged, and thus the apparently contrasting behavior of the Hall effect in lightly-doped LSCO and LNSCO can be compatible with the existence of the stripes in both systems. Another possible source of difference between the two systems is the particle-hole symmetry inside the stripes: It has been proposed that the vanishing Hall coefficient in LNSCO is essentially due to the particle-hole symmetry realized by the 1/4-filled nature of the stripes near the 1/8 doping;[26,27] if, on the other hand, the stripes at small x values are not exactly 1/4 filled, it is natural to observe non-vanishing Hall coefficient in LSCO, in the context of these theories.[26,27] Also, it is possible that the finite Hall resistivity in LSCO is caused the transverse sliding of the stripe as a whole; in fact, very recent optical conductivity measurements of lightly-doped LSCO have concluded that the sliding degrees of freedom are important for the realization of the metallic transport in this system.[28]

From the above discussion, it is clear that the metallic in-plane charge transport we observe in the AF state is most likely governed by the charge stripes. Given the fact that the hole mobility at moderate temperatures is surprisingly insensitive to the hole doping all the way up to optimum doping, it is tempting to conclude that the charge transport in cuprates that show the maximal T_c is also governed by the stripes. Recent STM studies of optimally-doped $Bi_2Sr_2CaCu_2O_{8+\delta}$ compounds, where periodic spacial modulations of the local density of state are observed,[29,30] also seem to support this conclusion. The implication of such a conclusion on our understanding of the high-T_c superconductivity is rather significant. Since the ordered static stripes are known to *kill* superconductivity, it must be the fluctuating nature of the stripes that facilitate the superconductivity at such high temperatures. There are already some theoretical proposals to explain the high-T_c superconductivity on the basis of the fluctuating stripes[14,31,32] or charge fluctuations.[33] The system we are dealing with may indeed be the "electronic liquid crystals",[15] which are quantum-fluctuating charge stripe states; our recent studies of the in-plane resistivity anisotropy of lightly-doped cuprates have found[21] that the resistivity is smaller along the stripe direction but the magnitude of the anisotropy is strongly dependent on temperature, which suggests a crossover between different electronic liquid crystal phases occurring in the cuprates, and the low-temperature phase appears to be an electron nematics.[34] Clearly, more experiments are needed to fully understand such a new state of matter, and to finally elucidate the mechanism of the high-T_c superconductivity.

5. Summary

It is shown that the doped holes in cuprates are surprisingly mobile in the long-range-ordered antiferromagnetic state at moderate temperatures, which is evidenced both by the metallic $\rho_{ab}(T)$ behavior and by the almost temperature-independent $R_H(T)$. It is emphasized that the *mobility* of the doped holes at moderate temperatures is virtually unchanged from the lightly hole-doped antiferromagnetic compositions (where the dominance of the stripes is very likely) to the optimally-doped superconducting composition, which implies that the charge transport even at optimum doping is essentially governed by the stripes.

Acknowledgments

This work was done in collaboration with A. N. Lavrov, S. Komiya, X. F. Sun, and K. Segawa. Stimulating discussions with S. A. Kivelson are greatly acknowledged. We also thank D. N. Basov, A. Fujimori, and J. M. Tranquada for collaborations and helpful discussions.

References

1. M. A. Kastner, B. J. Birgeneau, G. Shirane, and Y. Endoh, *Rev. Mod. Phys.* **70**, 89 (1998).
2. B. Keimer *et al.*, *Phys. Rev.* B **46**, 14034 (1992).
3. Y. Ando, A. N. Lavrov, and K. Segawa, *Phys. Rev. Lett.* **83**, 2813 (1999).
4. A. N. Lavrov, Y. Ando, K. Segawa, and J. Takeya, *Phys. Rev. Lett.* **83**, 1419 (1999).
5. T. Thio and A. Aharony, *Phys. Rev. Lett.* **73**, 894 (1994).
6. S. Komiya, Y. Ando, X. F. Sun, and A. N. Lavrov, *Phys. Rev.* B **65**, 214535 (2002).
7. Ch. Niedermayer *et al.*, *Phys. Rev. Lett.* **80**, 3843 (1998).
8. Y. Ando, A. N. Lavrov, S. Komiya, K. Segawa, and X. F. Sun, *Phys. Rev. Lett.* **87**, 017001 (2001).
9. Y. Ando, G. S. Boebinger, A. Passner, T. Kimura, and K. Kishio, *Phys. Rev. Lett.* **75**, 4662 (1995).
10. G. S. Boebinger, Y. Ando, A. Passner, T. Kimura, M. Okuya, J. Shimoyama, K. Kishio, K. Tamasaku, N. Ichikawa, and S. Uchida, *Phys. Rev. Lett.* **77**, 5417 (1996).
11. V. J. Emery and S. A. Kivelson, *Phys. Rev. Lett.* **74**, 3253 (1995).
12. T. Yoshida, X. J. Zhou, T. Sasagawa, W. L. Yang, P. V. Bogdanov, A. Lanzara, Z. Hussain, T. Mizokawa, A. Fujimori, H. Eisaki, Z.-X. Shen, T. Kakeshita, and S. Uchida, cond-mat/0206469.
13. E. Dagotto, *Rev. Mod. Phys.* **66**, 763 (1994).
14. E. W. Carlson, V. J. Emery, S. A. Kivelson, and D. Orgad, cond-mat/0206217.

15. S. A. Kivelson, E. Fradkin, and V. J. Emery, *Nature* **393**, 550 (1998).
16. J. Zaanen, *J. Phys. Chem. Solids* **59**, 1769 (1998).
17. J. M. Tranquada, B. J. Sternlieb, J. D. Axe, Y. Nakamura, and S. Uchida, *Nature* **375**, 561 (1995).
18. K. Yamada *et al.*, *Phys. Rev.* B **57**, 6165 (1998).
19. H. A. Mook, P. Dai, and F. Dogan, *Phys. Rev. Lett.* **88**, 097004 (2002).
20. A. W. Hunt, P. M. Singer, K. R. Thurber, and T. Imai, *Phys. Rev. Lett.* **82**, 4300 (1999).
21. Y. Ando, K. Segawa, S. Komiya, and A. N. Lavrov, *Phys. Rev. Lett.* **88**, 137005 (2002).
22. A. N. Lavrov, Y. Ando, S. Komiya, I. Tsukada, *Phys. Rev. Lett.* **87**, 017007 (2001).
23. T. Noda, H. Eisaki, and S. Uchida, *Science* **286**, 265 (1999).
24. J. M. Harris, Y. F. Yan, and N. P. Ong, *Phys. Rev.* B **46**, 14293 (1992).
25. M. Matsuda *et al.*, *Phys. Rev.* B **65**, 134515 (2002).
26. V. J. Emery, E. Fradkin, S. A. Kivelson, and T. C. Lubensky, *Phys. Rev. Lett.* **85**, 2160 (2000).
27. P. Prelovsek, T. Tohyama, and S. Maekawa, *Phys. Rev. B* **64**, 052512 (2001).
28. M. Dumm, D. N. Basov, S. Komiya, and Y. Ando, unpublished.
29. C. Howald, H. Eisaki, N. Kaneko, and A. Kapitulnik, cond-mat/0201546.
30. J. E. Hoffman, K. McElroy, D.-H. Lee, K. M. Lang, H. Eisaki, S. Uchida, and J. C. Davis, to be published in *Science*.
31. V. J. Emery, S. A. Kivelson, and O. Zachar, *Phys. Rev.* B **56**, 6120 (1997).
32. E. W. Carlson, D. Orgad, S. A. Kivelson, V. J. Emery, *Phys. Rev.* B **62**, 3422 (2000).
33. C. Castellani, C. Di Castro, and M. Grilli , *Phys. Rev. Lett.* **75**, 4650 (1995).
34. E. Fradkin, S. A. Kivelson, E. Manousakis, and K. Kho, *Phys. Rev. Lett.* **84**, 1982 (2000).

DYNAMIC INHOMOGENEITIES IN CUPRATES AND CHARGE-DENSITY WAVE SYSTEMS

D.MIHAILOVIC

Jozef Stefan Institute and Faculty of Mathematics and Physics,

University of Ljubljana,

Ljubljana, Slovenia

E-mail: dragan.mihailovic@ijs.si

Femtosecond time-resolved spectroscopy gives a unique viewpoint for examining the dynamics of elementary low-energy excitations in complex systems. Here we examine cuprates and charge-density wave systems (CDW), where the experiments point towards the existence of dynamically inhomogeneous states. Clear parallels can be drawn between precursor CDW formation and pairing inhomogeneities in cuprates. In the cuprates, the existence of a dynamically inhomogeneous state can help resolve a number of unresolved inconsistencies between time-resolved experiments, ARPES, STM and the existence of an anisotropic (*d*-wave) wave pairing interaction.

1. Introduction

The occurrence of dynamic inhomogeneities seems to be a ubiquitous feature of “complex matter”. These inhomogeneities may be anything from the formation of short-coherence length superconducting pairs or stripes in high-temperature superconductors (HTS) to segments of a charge density wave, nano-domains in relaxors or doping-induced inhomogeneities in transition metal dichalcogenides. The timescales on which these inhomogeneities occur also vary enormously, and often within the same system relaxation processes occur on timescales ranging from seconds to femtoseconds and everywhere in between. The experimental challenge of the last decade has been to invent and perfect new techniques for the investigation of dynamic inhomogeneities on time-scales sufficiently short to "freeze the motion" of the relevant excitations and give information on the microscopic origin of the dominant interactions leading to the observed inhomogeneity.

Recent experiments on cuprate superconductors have clearly shown the appearance of lattice inhomogeneities on relatively short timescales of the order of the inverse of the superconducting gap frequency ~ 10^{-13} s. Particularly pair-distribution function (PDF) and XAFS[1,2,3,4] studies, which have relatively short characteristic timescales reveal dynamic short-range correlations which are not observed with standard X-ray and neutron diffraction techniques. The results of these experiments appear to connect with inelastic neutron scattering[5], which shows anomalies on the length scale of the coherence length of 1 - 2 nm.

These experimental observations inhomogeneities are very important for the development of new theoretical models, which try to capture these features and make the connection between these inhomogeneities and the appearance of functional properties, such as pairing and superconductivity, or giant magnetoresistance for example.

Here we focus on a discussion of the quasiparticle (QP) relaxation dynamics in cuprates and charge-density wave systems on timescales of the order of 10^{-13} s, which we interpret from a viewpoint of dynamic inhomogeneities. In the last few years, femtosecond timescale time-resolved spectroscopy experiments have been performed on many HTS materials revealing quite a few common features, which cannot be understood in the framework of a homogeneous Fermi-liquid state. Similar experiments were also performed on charge-density-wave[6] systems revealing some close similarities between the pseudogap state in the cuprates and the precursor CDW state above the 3D CDW ordering temperature. We will briefly summarize some of these experiments and discuss the consequences in terms of an intrinsically inhomogeneous state.

2. Time - resolved measurements on HTS and CDW systems

Femtosecond spectroscopy involves the measurement of the photo-induced changes of transmission or reflection from a sample using laser pulses of the order of 70 fs or less.

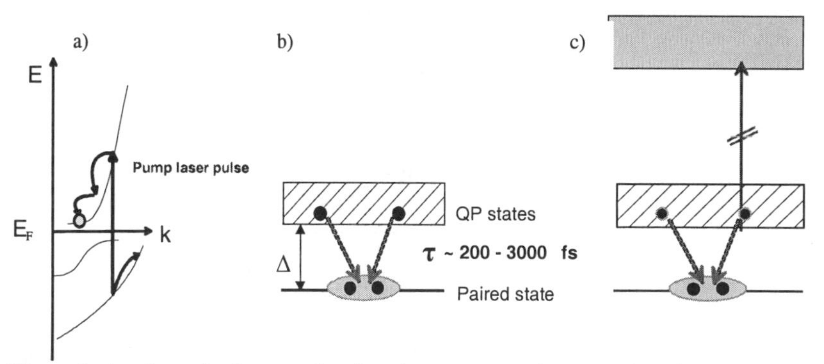

Figure 1. A schematic diagram showing the sequence of events after photoexcitation by an ultra-short laser pulse.

The experiments employ a pump-probe technique, which enables relaxation times to be measured directly. The sample is first excited with a laser pulse and is subsequently probed with a much weaker pulse of the same, or different wavelength.

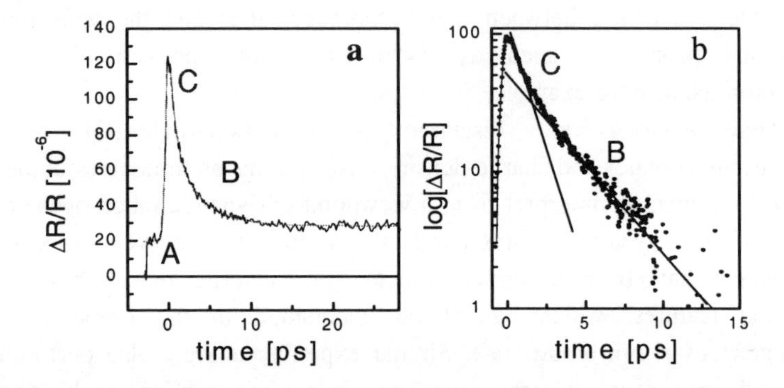

Figure 2 a) Raw data showing the three components typically observed in photoinduced transient reflectivity change in optimally doped YBCO. b) the same data on a log scale (with the background subtracted).

The pump pulse photons, which are absorbed within ~0.1 μm of the surface of the material, create a non-equilibrium population of electrons and holes which rapidly relax towards equilibrium by electron-electron scattering and scattering from phonons.

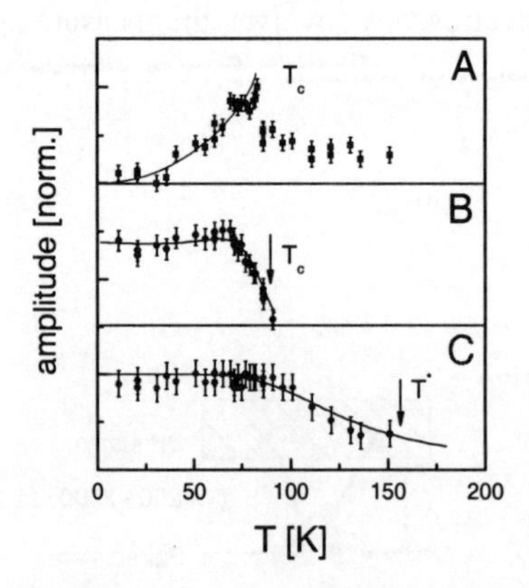

Figure 3. The T-dependence of the amplitude of the three components in the relaxation process in YBCO shown in Figure 2.

These initial relaxation processes are very rapid and in a metal photoexcited carriers reach the Fermi level within approximately 10~ 40 fs after excitation. However, if there is a gap in the density of states, such as might occur in a semiconductor, superconductor or a charge density wave system, the relaxation cascade is interrupted by the presence of the gap, and a bottleneck forms as carriers accumulate at the bottom of the band (see Figures 1 and 5 for a schematic representations of the relaxation scenario). The next relaxation step across the gap in a superconductor involves the creation of a superconducting pair from two quasiparticle states (Fig. 1). A similar process takes place in a CDW system, except that electron-hole recombination takes place. The population of the non-equilibrium QPs accumulated at the bottom of the band at any given time can be probed by a suitably delayed laser pulse, by a process resembling excited state absorption (see Fig. 1).

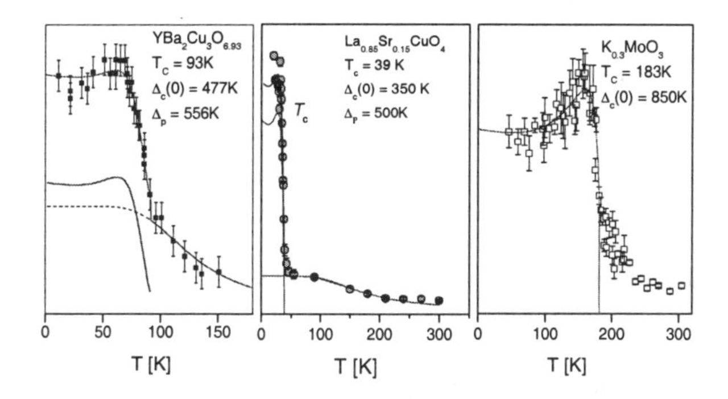

Figure 4. A comparison between the temperature dependence of the transient ampltide in YBCO, La2-xSrxCuO4 and KMO. The fits are made with the model of Kabanov et al. The value of the gap deduced from the fits are shown for both the pseudogap and the superconducting gap.

By now a large amount of data exist on femtosccond spectroscopy of YBCO[7,8,9,10,11,12,13,14,15,16,17], LaSrCuO[18,19], BiSCO[20,21,22], the various Tl[23] and Hg[24] compounds as well as electron-doped materials such as $Nd_{1.85}Ce_{0.15}CuO_{4-y}$ [25]. The results of experiments from different groups show very close similarities, and at present there are no major discrepancies between the data presented by different groups. The main features of the data which are relevant to the present discussion are shown in Figure 2 on optimally doped YBCO-123[11]. The time-response always has a number of components, which can - in virtually all cases - be separated into two timescales. The first is in the range 10^{-12} to 10^{-14} s, while the second is typically much longer.

The long timescale response (labeled A in Figure 2a)) appeears to be glass-like, having no well-defined lifetime, extending from 10 ps to over 100 μs. It usually appears as a uniform, slowly decaying background which builds up from pulse to pulse. Subtracting this background, an exponential plot of the raw data on YBCO-123 (Fig. 2b) reveals the presence of two fast-decaying components. One signal appears at the pseudogap temperature (around 150 K in optimally doped YBCO) and has a lifetime - which is temprature independent - of approximately 200-300 fs. The other component (B) is visible only below T_c and has a lifetime which is quite strongly temperature dependent. Importantly, as T_c is approached from below, the lifetime appears to diverge as $T \rightarrow T_c$ (Fig. 5).

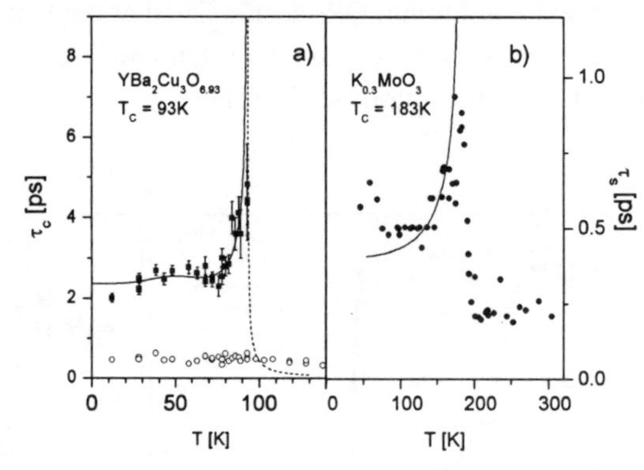

Figure 5. The relaxation times in optimally doped YBCO and KMO. Both show a divergence near T_c.

The amplitude of all three signals as a function of temperature is shown in Figure 3 for optimally doped YBCO. The sign of the two fast signals may vary from material to material and the probe wavelength. For example in YBCO-124[17], Tl-2223[23] etc. the signs of the two fast signals can be opposite with 800 nm probe pulses, while in Hg-1223[24], YBCO-123[10,11], LaSrCuO[19] and NdCeCuO[25] the two signals have the same sign response. The difference in sign is attributed to the anisotropy of the matrix elements in the optical probe transition between the accumulated QP states at the bottom of the band and the higher unoccupied states well above the Fermi level. It is therefore not directly relevant to the discussion of the low-energy gap structure. This is confirmed by the fact that the responses have the same T-dependence irrespective of the sign of the signal. The temperature dependence of the total amplitude of the fast transient signal is shown in Figure 4 for YBCO, LaSCO and the quasi-1D CDW chain compound $K_{0.3}MoO_3$ (KMO) [26]. The similarity of the response of the three systems is very characteristic and indicates the existence of common features in the carrier relaxation dynamics.

The relaxation time of the two fast relaxation components (B and C) for optimally doped YBCO is compared with KMO in Figure 5. In both cases, the relaxation process which appears below T_c shows a characteristic temperature dependence with an unmistakable divergence as T_c is approached from below.

3. Interpretation of time-resolved experiments within a homogeneous state model

A quantitative model has been developed by Kabanov et al.[10] to describe the temperature dependence of the amplitude and recombination time of the photoinduced signals. The model specifically addresses *s* and *d*-wave superconductors and compares the predicted behavior in the two types of systems. The model is also applicable to time-resolved spectroscopy in other gapped systems, since it does not distinguish between pair recombination and particle-hole recombination across the gap. It has been successfully applied to many cuprates, including YBCO-123[11] and $La_{2-x}Sr_xCuO_4$[19] over a wide range of doping, YBCO-124[17], Hg-1223[24] etc. When a uniform gap (such as an s-wave gap) is used in the model, quantitative fits to the model give remarkable agreement with the gap data from other spectroscopies, such as single particle tunneling for example. Such fits are shown in Figures 4 and 5 for the data on YBCO, LaSCO and KMO. However, repeated attempts to fit the data using a simple *d*-wave density of states have failed. In a *d*-wave scenario, after the fast initial e-e and e-p relaxation cascade (process 2 in Fig. 6), the QPs populate all *k*-vectors approximately equally. Since the overall DOS in a simple *d*-wave picture has a power-law density of states, no clear gap would then be evident[9]. Moreover, QP scattering from large gap regions to the nodal regions (process 4) would quickly deplete the large-gap regions of *k*-space leading to even more pronounced gapless relaxation, quite contrary to observations, which indicate relaxation across a well-formed large gap.

Since the experimental results were extremely reproducible and there was no question of sample surface degradation etc. (the optical measurements discussed here are not surface sensitive, since the light penetrates at least 100 nm into the sample.), the time-resolved data present a serious challenge to a homogeneous *d*-wave picture in the cuprates. In the light of the results of many other experiments showing a *d*-wave gap structure (particularly surface probes), clearly this apparent discrepancy needs to be understood.

It should be mentioned that alternative explanations of the QP recombination process have been discussed recently to try and circumvent this problem, attempting to make the results more consistent with a *d*-wave picture.

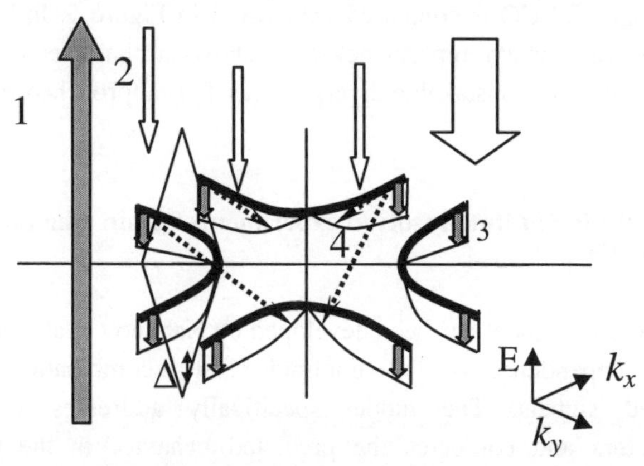

Figure 6. A 3D plot of the relaxation processes in time-resolved experiments following laser excitation (1). The initial e-e and e-ph relaxation processes (2) are very rapid. The QPs can relax by recombination (3) or scattering (4). The data indicate that the QP scattering process (4) into the nodal regions is not very relevant or does not occur.

One such scenario invokes a photoinduced transfer of spectral weight in the imaginary part of the optical conductivity, which would result in a transient change of the reflectance at optical frequencies around 1.5 eV[16]. This explanation - although appealing at first sight - does not describe the sign of the photoinduced signals correctly and is not consistent with the overall observations, predicting behavior which is contrary to what is observed. It cannot explain the T-dependence of the observed signals as shown in Figs 3-5, nor can it explain the multi-component relaxation behavior particularly above T_c. Thus a different, more elaborate explanation for the apparent discrepancy between the *d*-wave picture and the time-resolved experiments needs to be found.

4. A discussion of the importance of inhomogeneities

The time-resolved optical data summarized above can be reconciled with other experiments if we assume that the system is inhomogeneous on a timescale of the measurements (i.e. picosecond and sub-picosecond timescale). A parallel can be drawn between the behavior of CDW systems and cuprates, which gives valuable insight into the dynamically inhomogeneous state of the latter.

In the case of quasi-1D CDW materials such as $K_{0.3}MoO_3$, the appearance of dynamic inhomogeneities above T_c is relatively well understood. At room temperature, these materials are typically metals with a large anisotropy.

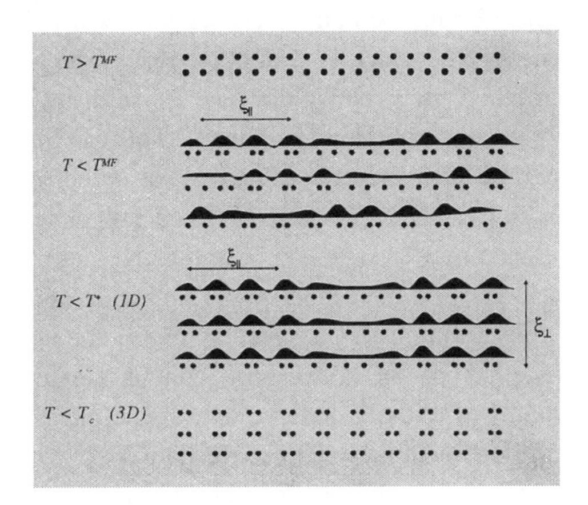

Figure 7. A graphical representaion of 1D CDWs at different temperatures. T_{MF} is the Mean-Field temperature. Adapted from Grüner[27])). At intermediate temperatures, the state is dynamically inhomogeneous.

On cooling, a sequence of orderings takes place (Fig. 7). First, as a result of a Peierls instability, short range CDW segments begin to appear on individual chains, resulting in a state which is both structurally and electronically inhomogeneous. The state is thought to be dynamically inhomogeneous because the CDW segments are not static, but fluctuate in time due to thermal fluctuations. The number of segments is statistically fluctuating and each segment represents a gapped region in real space, due to the local formation of the CDW. Eventually, at T_c, the chains acquire phase coherence and the inhomogeneity gradually disappears well below T_c.

In the time-resolved experiments (Ref. 26), the amplitude of the photoinduced signal is proportional to the total number of accumulated carriers above the gap and therefore directly proportional to the number of gapped CDW segments at any given temperature. The precursor inhomogeneous state with dynamically fluctuating CDW segments thus manifests itself as a precursor "pseudogap"-like feature above T_c in the amplitude of the time-resolved response in KMO (Fig. 4c).

In cuprates, instead of CDW segments, we propose a viewpoint by which a dynamically inhomogeneous state is formed by the presence of local superconducting pairs[28]. The associated inhomogeneity length scale is given by $l \sim \xi_s$ where ξ_s is the superconducting coherence length. The timescale of the dynamics associated with this type of inhomogeneity is determined by the fluctuations associated with excitations across the local pairing gap. The pairs have a binding energy 2Δ, which is a local property. The population of pairs at any given temperature T is given by the relevant statistics of a two-component system of bound pairs and unbound fermions[29]. This can be modeled rather well using a 2-level system, such as has been developed in the generalized bi-polaron model of Alexandrov et al.[30]. The time-resolved experiments, which show the unambiguous presence of a "pseudogap" can thus be understood in terms of the appearance of local gaps on the femtosecond timescale associated with pairing. The observed relaxation time is then simply the effective QP-to-pair recombination time across the local gap. This is single-valued and has no dispersion, explaining why the T-independent single-gap model of Kabanov et al.[9] gives such excellent quantitative fits to the data.

As temperature is lowered and/or the density of pairs increases with increasing hole doping, pairs start to interact with each other, aggregating into

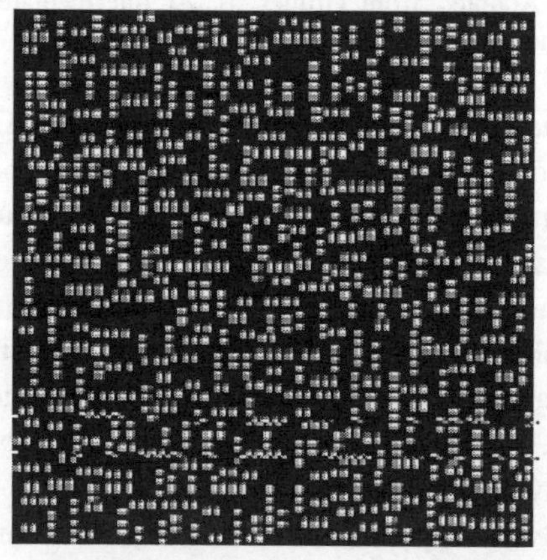

Figure 8. A real-space schematic representation of the CuO_2 planes including pairs and stripes of various lengths at a doping level corresponding to percolation threshold

larger objects, such as short stripes or clusters, leading to an increasingly complex inhomogeneous state (see Fig. 8). These objects are clearly highly anisotropic - as dictated by the symmetry of the interactions - and the formation of stripes is expected to lead to the formation of states for which the electron momentum *k* becomes increasingly well defined as a quantum number. Symmetry analysis shows that the formation of stripes is allowed both along the bond axes and along the diagonals and both types of objects may be expected to appear.

The timescale associated with the population fluctuations of these larger objects is expected to be slower than for the pair recombination. The intra-gap excitations with glass-like relaxation dynamics which are ubiquitously observed in femtosecond time-resolved experiments in cuprates[24] may thus be associated with the dynamics of a complex inhomogeneous state including various stripes and clusters.

With increasing doping, as inter-pair interactions become unavoidable, collective effects become important. In KMO the appearance of a 3D correlated CDW state is clearly indicated by the appearance of a T-dependent BCS-like gap in both the amplitude of the response (Fig. 4c) and the lifetime (Fig.5). The observation of a collective T-dependent gap in the time-resolved measurements (as indicated by the divergence of recombination time below T_c and the strong temperature-dependence of the amplitude near T_c.) is a clear a manifestation of such collective phenomena in cuprates as well. However, the *coexistence* of two relaxation times such as ubiquitously observed in cuprates, one associated with pairing, and the other with collective excitations may be interpreted as further evidence that the state is inhomogeneous.

Finally, commenting on some other experimental data, the appearance of states crossing the Fermi energy in some directions together with a "pseudogap" in other direction in ARPES experiments[31,32,33,34] may be interpreted quite consistently as the appearance of an inhomogeneous state. Metallic (stripe) states for which *k* is reasonably well defined along certain directions would result in broken “Fermi-liquid-like” sections, together with a pseudogap elsewhere. Since ARPES presents a map of electron momenta leaving the sample in a way that is spatially averaged over the surface of the sample, the spectrum necessarily represents the sum of all the different contributions, and is thus an inhomogeneous average over the surface. The usual appearance of a superconducting temperature-dependent gap (including the peak-dip-hump structure) along certain directions together with a normal state gap can thus be made fully consistent with the proposed scenario of an inhomogeneous state deduced on the basis of femtosecond time-resolved experiments.

Scanning tunneling microscopy (STM), which measures only the surface layers, just like ARPES shows similar features and can be interpreted in a similar fashion[35,36,37,38]. The difference is that STM maps of the local DOS are averaged over *k*. Nevertheless, experiments on YBCO and BiSCO have shown that the surface of these materials is highly inhomogeneous. There are large variations in the local shape of the density of states on the surface which can be interpreted either at face value - where they show variations in the DOS arising from a charge inhomogeneity - or as evidence for quasiparticle scattering by impurities arising only at special points on the FS. (The two viewpoints are not necessarily unreconcilable.) Yet both clearly show that the typical scattering length in cuprates is very small. The peaks in the STM Fourier transforms are near ($\pi/2a$), indicating a repeat period of approximately one coherence length, or the inter-particle distance in a mesoscopic Jahn-Teller pair[28].

A more serious issue of fundamental importance with STM experiments and ARPES is whether the inhomogeneities which are observed as static exist also in the bulk, as they clearly are on the surface. The combined evidence from standard diffraction studies (where short-range inhomogeneities are not clearly observed) and PDF probes – (where they are), suggests that the bulk may not have static inhomogeneities, while the time-resolved experiments appear to suggest that the bulk is dynamically inhomogeneous in a similar way that the precursor CDW state is inhomogeneous above T_c.

5. Conclusions

It is becoming increasingly clear that a great deal of apparently controversial experimental data can be reconciled by moving away from homogeneous state models in the cuprates and considering inhomogeneity as an intrinsic and essential feature of therese materials. This inhomogeneity may be be viewed ultimately to be the cause of the observed functional properties, including supercondutivity in HTS. It appears that in may cases new theoretical approaches are necessary to decribe such behaviour[39,28,40,41, 42,43,44,]

Acknowledgments

I wish to thank V.V.Kabanov for critical reading and D.Dvorsek for help in preparation of the MS.

References

1. T.Egami and S..J.L.Billinge, in "Physical properties of High-Temperature Superconductors", Ed. D.M.Ginsberg (World Scientific, Singapore, 1996), p.265, E.S.Bozin *et al*, Phys.Rev.Lett. **84**, 5856 (2000).
2. A. Bianconi *et al.*, Phys. Rev. Lett. **76,** 3412 (1996), N. L. Saini *et al.*, Phys. Rev. B **55**, 12759 (1997).

3. D. Haskel *et al.*, Phys. Rev. B **61**, 7055 (2000) .
4. S. D. Conradson, J. M. DeLeon, A. R. Bishop, J. Supercond. **10,** 329 (1997).
5 R.J.McQueeney *et al*, Phys. Rev. Lett, **82**, 628 (1999).
6 J. Demsar *et al.*, Phys. Rev. B **66** art. no. 041101 (2002) .
7 S.G.Han *et al.*, Phys.Rev.Lett. **65**, 2708 (1990).
8 D.H.Reitze *et al.*, Phys. Rev. B **46**, 14309 (1992).
9 C.J.Stevens *et al.*, Phys.Rev.Lett. **78**, 2212 (1997).
10 V.V. Kabanov *et al.*, Phys. Rev. B **59**, 1497 (1999).
11 J. Demsar *et al.*, Phys. Rev. Lett. **82**, 4918 (1999).
12 O.V. Misochko, K. Sakai, S. Nakashima, Physica C **329**, 12 (2000).
13 R.A. Kaindl *et al.*, Science **287**, 470 (2000).
14 P. Gay et al., Physica C **341-348**, 2269 (2000).
15 R.D. Averitt *et al.*, Phys. Rev. B **63**, 140502 (2001).
16 G.P. Segre *et al.*, Phys.Rev.Lett. **88**, 137001 (2001).
17 D. Dvorsek *et al.*, Phys. Rev. B **66**, art. no. 020510 (2002).
18 S. Rast *et al.*, Phys. Rev. B **64**, 214505 (2001)
19 P. Kusar, unpublished.
20 J. M. Chwalek *et al.,* Appl. Phys. Lett. **57**, 1695 (1990).
21 D. C. Smith *et al.*, Physica C **341-348**, 2269 (2000).
22 H. Murakami *et al.*, Physica C **367**, 317 (2002).
23 G. L. Eesley *et al.*, Phys. Rev. Lett. **65**, 3445 (1990); D. C. Smith *et al.*, Physica C **341**, 2219 (2000).
24 J. Demsar *et al.,* Phys. Rev. B **63**, art. no. 054519 (2001).
25 L. Yongqian *et al.*, Appl. Phys. Lett. **63**, 979 (1993).
26 J. Demsar, K. Biljakovic, D. Mihailovic, Phys. Rev. Lett. **83**, 800 (1999).
27 G. Grüner, *Density Waves in Solids* (Addison-Wesley, Reading, MA, 1994).
28 D. Mihailovic, V. V. Kabanov, Phys. Rev. B **63**, art. no. 054505 (2001).
29 V. V. Kabanov, D. Mihailovic, Phys. Rev. B **65** art. no. 212508 (2002).
30 A.S.Alexandrov and N.F. Mott, *High Temperature Superconductors and other Superfluids* , (Taylor and Francis , 1994).
31 A. Damascelli, D. H. Lu, Z. X. Shen, J. Electronic Spectrosc **117** 165 (2001).
32 Y. D. Chuang *et al.*, Physica C **341**, 2079 (2000).
33 H. Ding *et al.*, Nature **382**, 51 (1996).
34 T. Sato *et al.*, Physica C **341**, 2091 (2000).
35 S. H. Pan *et al.*, Nature **413**. 282 (2001).
36 J. E. Hoffman *et al.*, Science **295**, 466 (2002).
37 J. E. Hoffman *et al.*, Science **297**, 1148 (2002).
38 A. de Lozanne, Supercond. Sci. Tech. **12**, R43 (1999).
39 L.P.Gorkov and A.B.Sokol, Pis'ma ZhETF, **46**, 333 (1987), JETP Lett. **46**, 420 (1987).
40 A. R. Bishop, D. Mihailovic, J. Mustre de Leon, unpublished.
41 A. Bussman-Holder *et al.*, J. Phys. Conds. Matt. **13**, L545 (2001).

42 S. R. Shenoy, V. Subrahmanyam, A. R. Bishop, Phys. Rev. Lett. **79**, 4657 (1997)
43 Y. N. Ovchinnikov, S. A. Wolf, V. Z. Kresin, Phys. Rev. B **63**, art. no.064524 (2001).
44 J. Mustre de Leon *et al.*, Phys. Rev. Lett. **68**, 3236 (1992)

IS THERE A NARROW CONDUCTIVITY MODE IN THE SUPERCONDUCTING OXIDES?

S.SRIDHAR, C.KUSKO AND Z.ZHAI

Department of Physics, Northeastern University, Boston, MA02215

Several anomalous electrodynamic results in the high temperature superconducting cuprate oxides are shown to be explained by the presence of a narrow conductivity mode (width $\Gamma < 0.1meV$). The presence of this mode provides a coherent description of the microwave electrodynamic response in both the superconducting and non-superconducting states of the cuprate superconductors. The features explained include the conductivity peaks observed in the superconducting state, and the signatures of a "microwave plasmon" in the pseudogap state.

The transition metal oxides display an astonishingly diverse range of phenomena: high temperature superconductivity, colossal magnetoresistance, ferroelectricity, spin gaps and charge density waves. A systematic set of microwave measurements on several oxides - cuprate superconductors, CMR manganites and striped nickelates - has yielded new insights into their electrodynamic properties.

In this paper we discuss some common features of the response of the oxides to electromagnetic radiation at microwave frequencies. In Fig. 1 we present the generic phase diagram of the high temperature superconducting cuprates (HTS) illustrating the various phenomena (charge fluctuations, spin glass, dielectric transitions, conductivity mode) observed in microwave experiments on the insulating, "normal" and superconducting states. A cohesive picture is emerging that the electrodynamics at these sub-optical frequencies is dominated by collective charge transport that can be described in terms of a narrow conductivity mode ($\Gamma < 0.1meV$).

In the superconducting state below T_c of all the cuprates, including $YBa_2Cu_3O_{7-\delta}$, $HgBa_2Ca_2Cu_3O_{8+\delta}$ (Hg : 1223), $HgBa_2CuO_{4+\delta}$ (Hg : 1201), and $Tl_2Ba_2CuO_{6+\delta}$ (Tl : 2201), the microwave data are clearly incompatible with the quasiparticle response of a pure d-wave superconductor. Instead, we have successfully modeled the data in terms of a non-quasiparticle scenario, according to which the electrodynamic response is

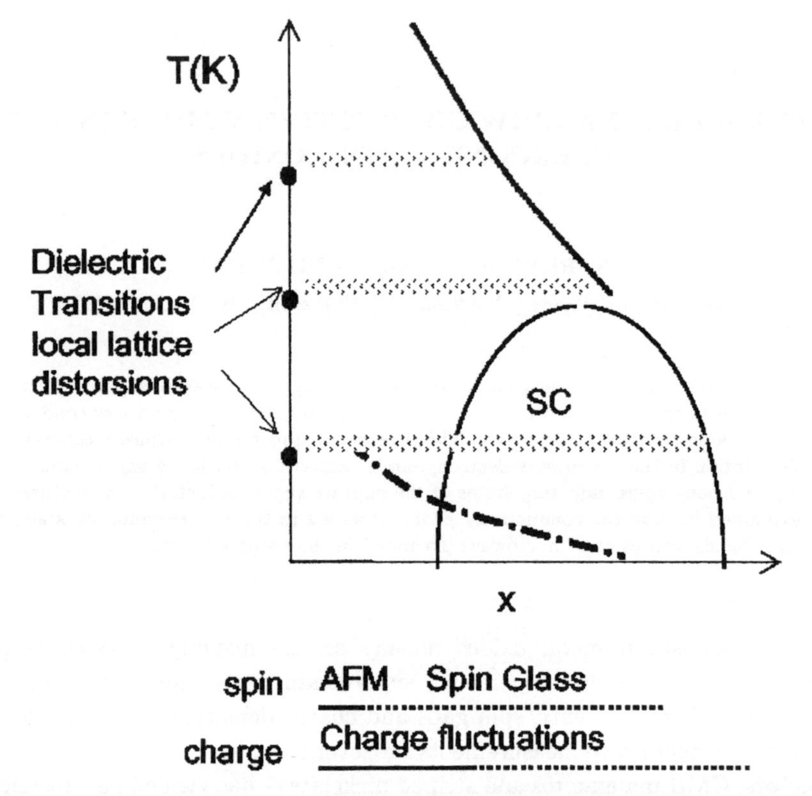

Figure 1. Phase diagram from the perspective of microwave measurements at 10 GHz. The principal features discussed are shown in the diagram

ascribed to a collective mode, similar to the dynamic response of the phason mode of density wave in low dimensional materials [1]. Such a density wave can well arise from an inhomogeneous electronic state. The conductivity mode dominates the underlying quasiparticle contribution.

The collective conductivity mode dynamics is also observed above T_c , in terms of an anomalous plasmon-like response implying negative permittivities ($\Re(\varepsilon(\omega)) < 0$) at microwave frequencies in the metallic pseudogap state of the cuprate superconductors. The microwave plasmon arises from a charge collective mode characterized by a low plasma frequency and extremely low damping, distinctly different from those observed at optical frequencies.

Another feature common to several oxides, including the parent members of the cuprate superconductors and the nickelates, is the observation of dielectric transitions occurring at characteristic temperatures that we associate with lattice instabilities. In La_2CuO_{4+x} and $La_{5/3}Sr_{1/3}NiO_4$ these occur at common temperatures $32K$ and $245K$, and are signatures of local lattice octahedra instabilities in these isostructural perovskite oxides. Similar dielectric transitions are also observed in non-superconducting $YBa_2Cu_3O_{6.0}$ at $65K$ and $110K$ - the latter correlates well with other reports of changes in the buckling angle and NQR reports of CDW formation. The microwave results are correlated with a variety of other measurements and reveal new aspects of the phase diagram of the perovskite cuprates and nickelates. These results also indicate that inhomogeneous electronic states, such as charge stripes and oxygen ordering, are strongly connected to underlying lattice instabilities.

1. Experimental Details

All the materials studied were in single crystal form [2]. Ultra-pure single crystals of $YBa_2Cu_3O_{6+x}$ with varying O concentration $x = 0+, 0.38, 0.51, 0.78, 0.86, 0.95, 1.0$ were prepared in contamination-free $BaZrO_3$ crucibles. The O concentrations cover the entire range from insulating, non-superconducting material ($x = 0+$) to optimum ($x = 0.95$) and over-doped ($x = 1.0$). The high quality of these single crystals has been extensively documented in a wide range of measurements, including structural and transport studies of the superconducting and non-superconducting states. Single crystals of $Hg : 1201$ ($T_c = 94.4K$), $Hg : 1223$ ($T_c = 122K$) and $Tl : 2201$ ($T_c = 91K$) were prepared by appropriate methods for each material. Single crystals of $La_2CuO_{4+\delta}$ and $La_{5/3}Sr_{1/3}NiO_4$ were prepared by the TSFZ method.

The precision microwave measurements were carried out in a Nb superconducting cavity resonant at $10GHz$ in the $TE011$ mode. The sample is placed at the center of the cavity at a maximum of the microwave magnetic field H_ω. A detailed analysis of the relevant cavity perturbation for general sample conditions including lossy dielectric and metallic or superconducting states, has been recently carried out by us [3]. We are able to directly measure the conductivity $\tilde{\sigma} - i\omega\varepsilon_0\tilde{\varepsilon}$. We have carried out extensive measurements of the surface impedance of a variety of superconductors, and demonstrated the validity of these measurements. The use of the superconducting cavity leads to extremely high sensitivity to the measurements of surface impedance, microwave conductivities and dielectric permittivities.

An important issue that has emerged from our analysis is that microwaves couple dominantly to the charge system, and negligibly to the spin system. Of course the charge system is also coupled to the spins and lattice, and hence in addition to charge instabilities, instabilities in the spin and lattice sub-systems show up also in the microwave response as illustrated in Fig. 2.

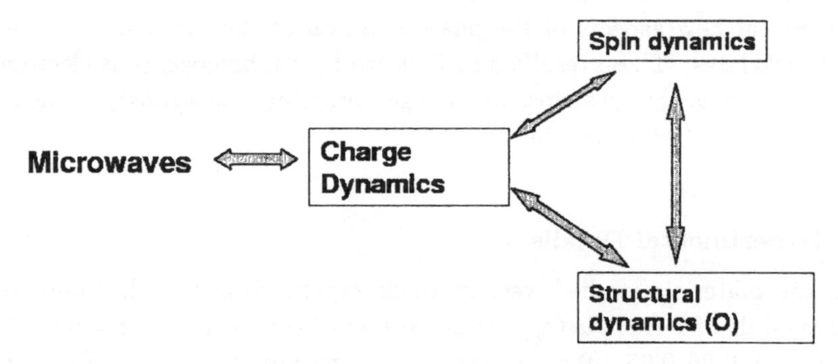

Figure 2. The coupling of the microwave field with the charge, spin and lattice degrees of freedom.

2. Superconducting state ($T < T_c$)

Just as the mechanism of superconductivity is unknown, an understanding of the mechanism responsible for the microwave response poses some important challenges. While many other experiments are consistent with a d-wave superconducting state, the microwave response, particularly the absorption, is inconsistent with a quasiparticle scenario that is based upon a BCS like d-wave density of states. Instead the quasiparticle response is dominated by a conductivity collective mode, as we show below.

2.1. Failure of the quasiparticle scenario in HTS

We have conclusively shown that the quasiparticle scenario fails for microwave measurements in the superconducting state of Bi : 2212 [4]. It should be emphasized that Bi : 2212 is not unique among the HTS - the same inconsistencies also occur in $YBa_2Cu_3O_{7-\delta}$ as we discuss below.

Below T_c, the scattering rate can be obtained from the complex conductivity $\tilde{\sigma} = \sigma_1 + i\sigma_2$. In a phenomenological "two-fluid" model the high frequency conductivity $\tilde{\sigma}(\omega, T)$ can be expressed as:

$$\tilde{\sigma}(\omega,T)=\sigma_1+i\sigma_2=\frac{ne^2}{m}\left[\frac{n_{qp}(T)}{-i\omega+1/\tau(T)}+\frac{in_s(T)}{\omega}\right] \tag{1}$$

where n_{qp} and n_s represent the fractions of normal and superconducting quasiparticles (with $n_{qp}(T)+n_s(T)=1$), and τ is the relaxation time for the normal electrons. In this model, the normal electrons are damped, leading to the usual Drude conductivity at high frequencies, and the superconducting electrons have inertia but no damping. In the local London limit, defined by the condition $\xi \ll \ell \ll \lambda$ which is well satisfied by $YBa_2Cu_3O_{7-\delta}$, n_{qp} and n_s can be calculated using the Mattis-Bardeen expression [5] : $n_s=(1-n_n)=1-2\left\langle\int_0^\infty\left(-\frac{\partial f}{\partial E}\right)d\epsilon\right\rangle$ where the quasiparticle energy is $E=\sqrt{\epsilon^2+\Delta_k^2}$ and Δ_k is the superconducting order parameter with a d-wave symmetry. We assume a BCS temperature dependence for the $d_{x^2-y^2}$ gap parameter $\Delta(T,\phi)=\Delta_d(T)\cos(2\phi)$. The d-wave model describes well the low T behavior of $\lambda(T)\propto T$ for $Bi_2Sr_2CaCu_2O_{8+\delta}$ $(Bi:2212)$[6] and $YBa_2Cu_3O_{6.95}$.

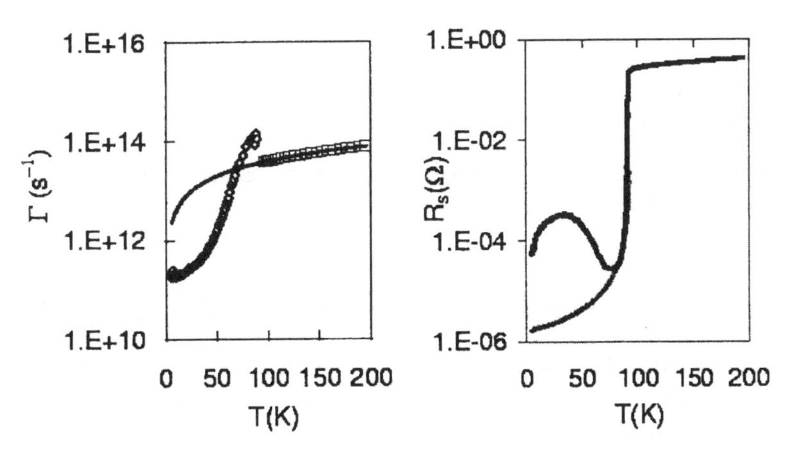

Figure 3. Left panel: scattering rate of $YBa_2Cu_3O_{7-\delta}$ deduced using a d-wave quasiparticle analysis. Right panel: microwave absorption R_s deduced from a linear extrapolation of Γ_{qp} (line) compared with experimental data (points).

From Eq. 1 we can calculate the inverse lifetime in the superconducting state using the approximation that the frequency ω is negligible in comparison with the scattering rate. In Fig. 3 we show the scattering rate $\Gamma(T)$ for $YBa_2Cu_3O_{6.95}$ calculated using a d-wave symmetry model. $\Gamma(T<T_c)=\tau^{-1}(T)$ drops rapidly at T_c and continues to decrease rapidly

at lower temperatures. An even larger reduction in $\Gamma(T < T_c)$ is required in $Bi : 2212$. This very large variation of Γ is necessary to quantitatively describe the large values of the microwave absorption within the quasiparticle framework. Within the quasiparticle scenario, if R_s and consequently σ_1 is large at low T, $\Gamma = \tau^{-1}$ has to be very small since $n_{qp}(T \to 0) \to 0$. (Even quasiparticle localization arguments are insufficient to explain the large σ_1 required).

Conversely, if one uses a reasonable (eg. linear) extrapolation of $\Gamma(T < T_c)$ then the microwave absorption calculated in the d-wave scenario is too low to explain the observed data. *Thus the gap-quasiparticle scenario fails to explain the large absorption observed in the microwave response of HTS.* This is a major problem not only for the microwave applications, but also for fundamental understanding of HTS.

The essential reason the quasiparticle scenario fails is because the data are inconsistent with the sum rule required for superconductivity, viz. $n_s(T) + n_{qp}(T) = 1$. This sum rule requires that $\sigma_1(T < T_c) \ll \sigma(T_c)$. This condition is clearly not satisfied in the HTS.

2.2. Conductivity mode explanation of the microwave data for $T < T_c$

We show that addition of a parallel conductivity channel, whose origin we ascribe to a collective mode, quantitatively describes the HTS data [4].

$$\sigma_{tot}(\omega, T) = \sigma_s(\omega, T) + \sigma_{CM}(\omega, T) = \sigma_s(T) + \frac{\sigma_{CM}(T)}{1 - i\omega\tau_{CM}(T)}$$

where we use the d-wave expression for σ_s from eq.1, and a Drude type expression for the collective mode conductivity σ_{CM}. Note that this is a limit case form of the more general expression

$$\sigma_{CM}(\omega, T) = \sigma_{CM}(T) \frac{-i\omega}{\tau_{CM}(\omega_{pin}^2 - \omega^2) - i\omega}$$
$$\approx \frac{\sigma_{CM}(T)}{1 - i\omega\tau_{CM}(T)} \qquad (\omega_{pin} \to 0) \qquad (2)$$

The frequency dependence used above for σ_{CM} has been well-established in the 1-D Charge Density Wave materials like $NbSe_3$and TaS_3 [1].

In the following we calculate the conductivity σ_1 due to the collective mode for $Hg : 1223$ and $YBa_2Cu_3O_{6.5}$ as illustrated in Fig. 4 and Fig.

5. Excellent fits to the data for these HTS in the superconducting state are obtained with the following functional dependences for the oscillator strength $\sigma_{CM}(T)$ and the scattering rate $\Gamma_{CM}(T)$:

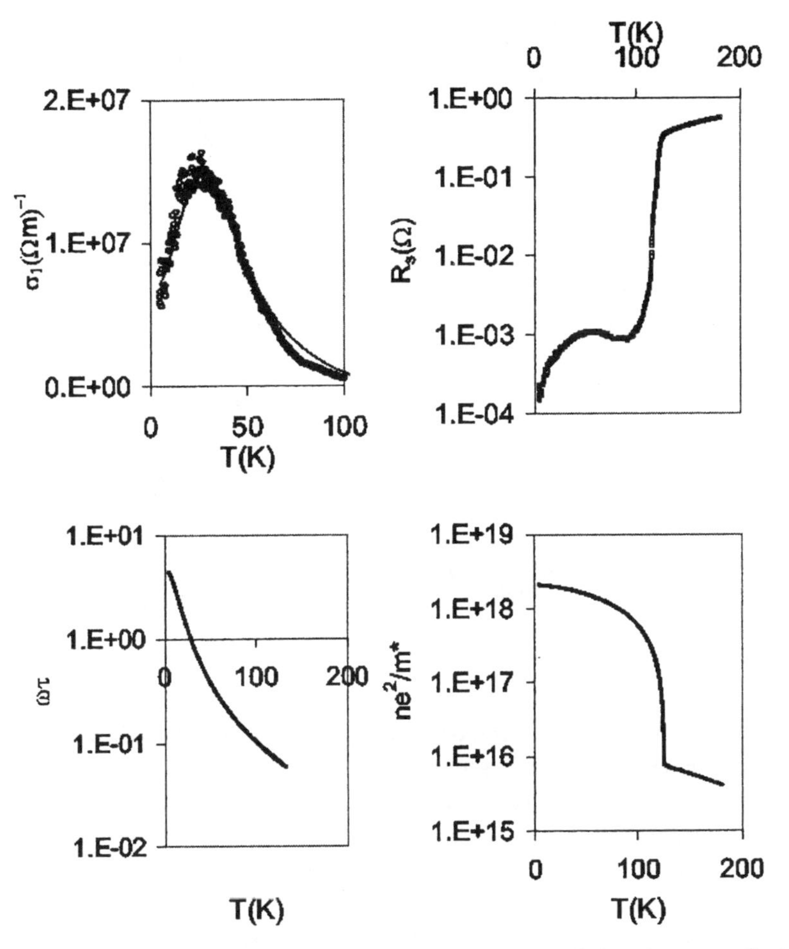

Figure 4. The fits of the microwave conductivity experimental data in terms of a low energy collective mode for Hg : 1223. Upper panels: (left) - the experimental microwave conductivity (empty circles) and the calculated microwave conductivity (solid line), (right) - the measured surface resistance R_s. Lower panels: (left) - the temperature dependence o $\omega\tau$, (right) - the temperature dependence of the oscillator strength.

(1) The oscillator strength of the collective mode increases linearly in both cases as the temperature decreases (see Fig.4 and 5), in sharp

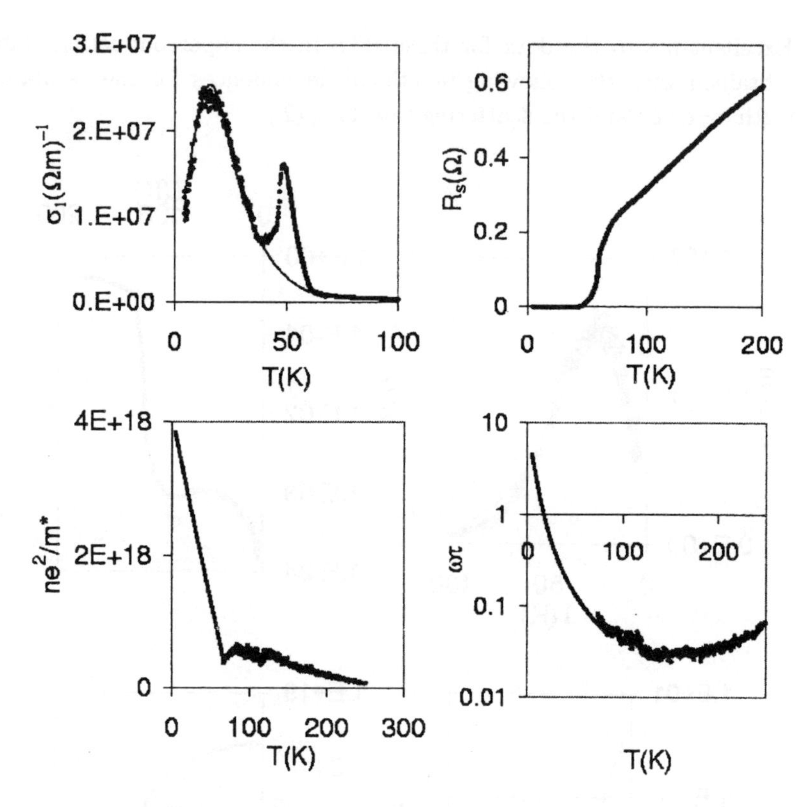

Figure 5. The same as Fig. 4 for $YBa_2Cu_3O_{6.5}$

contrast with the behavior of the oscillator strength used in the d-wave quasiparticle scenario.

(2) For $Hg:1223$ the scattering rate decreases monotonically with the temperature according to the relationship: $\Gamma_{CM}(T) = \tau_{CM}^{-1}(T) = (0.13+6\cdot10^{-4}T^2)\cdot10^{11}(sec^{-1})$. For $YBa_2Cu_3O_{6.5}$ the temperature dependence of the scattering rate has a similar behavior given by: $\Gamma_{CM}(T) = (0.15+6.6\cdot10^{-6}T^{4.2})\cdot10^{11}(sec^{-1})$.

We note that in the analysis $\sigma_2(T)$ is still described by the d-wave analysis using conventional values of $\lambda(T=0)$, while $\sigma_1(T)$ is described by the conductivity collective mode. However $\lambda(T=0)$ is not directly measured in any microwave experiment and are instead deduced from other experiments. Indeed if the accepted $\lambda(T=0)$ values do not apply at microwave frequencies, then the $\sigma_2(T)$ data is also explained by the collective mode. It is quite possible therefore that the linear T dependence of the penetration depth, cited as a signature of d-wave superconductivity, may well be due to the collective conductivity mode also.

Note that $\Gamma_{CM}(T)$ is of the same order of magnitude as the microwave frequency indicating the low energy scale of the collective mode. When the scattering rate is equal to the external microwave frequency $\omega/\Gamma = 1$ a peak in the conductivity appears. This conductivity peak is similar in spirit to the peak observed in internal friction studies, and dielectric loss peaks in glasses.

2.2.1. *Conductivity peaks*

An important feature of the collective mode explanation is the quantitative description of the conductivity peaks that are observed in the HTS cuprates. In the collective mode scenario, a peak in $\sigma_1(T)$ and hence $R_s(T)$ occurs at a temperature T_p where $\omega\tau_{CM}(T_p) = 1$. In $YBa_2Cu_3O_{7-\delta}$ and $HgBa_2Ca_2Cu_3O_y$, $T_p \sim 30K$, and a clear peak is observed. In contrast in $Bi : 2212$ $\omega\tau_{CM}(T) \sim 10^{-2}$ at all temperatures and hence no peak is observed. A similar situation also occurs in $YBa_2Cu_3O_{6.5}$. (See later for a discussion of multiple conductivity peaks).

A comparison of the deduced scattering time for several cuprates is shown in Fig.6. Thus within this scenario, the only difference among the cuprates is a quantitative one of different scattering rates, and clearly shows that speculations that $YBa_2Cu_3O_{7-\delta}$ is different from Bi:2212 are unfounded.

In the quasiparticle scenario, the conductivity peak is described as a crossover from a high T region where the increase of $\tau_{qp}(T)$ dominates to a low T region where the decrease of $n_{qp}(T)$ dominates [8]. In the collective mode scenario, the peak represents a crossover from (i) increasing $\sigma_{CM}(T)$ which dominates at high T where $\omega\tau_{CM}(T) \ll 1$ and $\sigma_{CM}(\omega, T) \sim \sigma_{CM}(T)$, to (ii) a low T region, where increasing $\tau_{CM}(T)$ dominates because $\omega\tau_{CM}(T) \gg 1$and $\sigma_{CM}(\omega, T) \sim \sigma_{CM}(T)/(\omega\tau_{CM}(T))^2$.

3. Anomalous response in the pseudogap state

Low energy (microwave) measurements of the surface impedance $Z_s = R_s + iX_s$ on $HgBa_2Ca_2Cu_3O_{8+\delta}$ $(Hg : 1223)$, $HgBa_2CuO_{4+\delta}$ $(Hg : 1201)$, $Tl_2Ba_2CuO_{6+\delta}$ $(Tl : 2201)$ and underdoped $YBa_2Cu_3O_{6.5}$ reveal new features of transport in the pseudogap state. A particularly striking observation is the deduction that in the "normal" state above T_c, the electrodynamic response is distinctly different from that of a conventional metal,

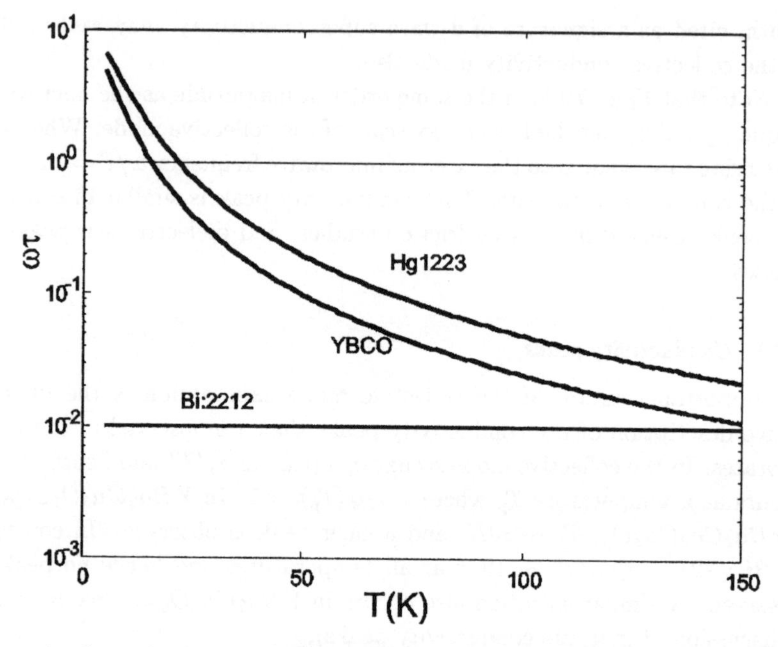

Figure 6. Scattering time of the collective mode $\omega\tau_{CM}(T)$ for several cuprates deduced from microwave measurements $\omega = 2\pi 10^{10} sec^{-1}$.

and can instead be characterized as a "microwave plasmon". The principal experimental observations that suggest this conclusion are two fold: (i) Above T_c the curves of R_s vs. T and X_s vs. T are not parallel, so that $R_s(T > T_c) \neq X_s(T > T_c)$! (ii) Furthermore $\Delta X_s(T_c) > \Delta R_s(T_c)$, exactly opposite to that observed in conventional metals like Nb and Sn.

Essentially similar data were found for all the materials in this study : $Hg : 1223$, $Hg : 1201$, $Tl : 2201$ and $YBa_2Cu_3O_{6.5}$ as illustrated in Fig.7.

The inequality $R_s(T > T_c) \neq X_s(T > T_c)$ was observed for all the 4 materials in terms of the anomaly $\mathcal{A} = X_s/R_s - 1$ vs. T . The anomaly $\mathcal{A}$ is clearly finite (non-zero) for $T > T_c$ for both orientations $H_\omega \parallel c$ and $H_\omega \perp c$. In optimally doped $YBa_2Cu_3O_{6.95}$ and $Bi_2Sr_2CaCu_2O_8$ the anomaly $\mathcal{A}(T > T_c)$ is smaller than the uncertainties in the experiment, but may well be finite.

The measurements indicate a breakdown of the so-called Hagen-Rubens limit (where the measurement frequency $\omega \ll \Gamma$, the carrier relaxation or dissipation rate), indicating plasmon-like response characterized by negative microwave dielectric permittivities $\varepsilon'(\omega) < 0$, for currents in the ab-plane. Such anomalous conduction in the pseudogap state indicates non-

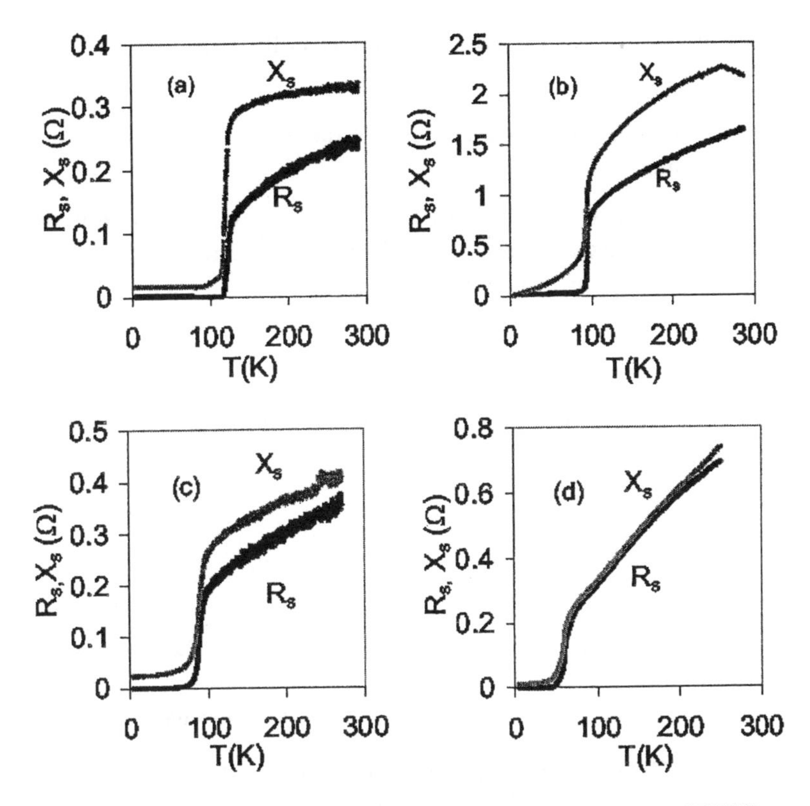

Figure 7. Surface resistance and surface reactance for a series of HTSC

Fermi Liquid (NFL) behavior rather than a single particle (Fermi liquid) transport mechanism, and that the microwave dynamics and the optical response are characterized by vastly different energy scales.

The anomalous microwave results are suggestive of a phason mode of a density wave, whose electrodynamic response can be represented as in Eq. 2 [1]. In the unpinned case $\omega \gg \omega_{pin} (\rightarrow 0)$, the response reduces to the Drude form $\tilde{\sigma}_{CM}(\omega) = \sigma_0/(1 - i\omega\tau_{CM})$ used above. A model based upon a collective phason mode arising from the presence of charge fluctuations, such as from stripes or a Density Wave (DW), quantitatively explains the observed temperature dependence of experimental data [11]. *This result confirms that even at very high temperatures the electrodynamics response is characterized by a narrow conductivity mode with $\omega \sim \Gamma = 0.1meV$.* The results are quantitatively explainable in terms of a collective phason mode. Such a phason mode response can arise from a DW order parameter [9], or also from stripe fluctuations, which have CDW-like dynamics [10].

4. Significant temperature scales: signatures of local lattice instabilities

Another strikingly common feature of the microwave response of the cuprates is the observation of multiple dielectric transitions at specific temperatures in the parent insulating compounds.

In $YBa_2Cu_3O_{6.5}$, $La_2CuO_{4+\delta}$ and also $La_{5/3}Sr_{1/3}NiO_4$, the dielectric permittivity over a wide range of temperature can be very well analyzed in terms of multiple dielectric modes, $\tilde{\varepsilon}(T) = \tilde{\varepsilon}_\alpha(T) + \tilde{\varepsilon}_\beta(T) + \tilde{\varepsilon}_\gamma(T) + ...$, each of which is well described by a Debye relaxation form with respect to the temperature dependence

$$\tilde{\varepsilon}(T) = \tilde{\varepsilon}_\alpha + \tilde{\varepsilon}_\beta + ... = \sum_{i=\alpha,\beta,\gamma,..} \frac{\varepsilon_{i0}(T)}{1 - i\omega\tau_i(T)}, \tag{3}$$

where $\varepsilon_{i0}(T)$ are static dielectric functions and $\tau_i(T)$ relaxation times for each mode. A key observation is that dielectric modes turn on at specific temperatures T_{di}, i.e. $\varepsilon_{i0}(T) \sim \varepsilon_{io}(0)(1 - T/T_{di})$, as discussed below.

(1) In $La_2CuO_{4.0175}$ three contributions can be identified. $\tilde{\varepsilon}_\alpha(T)$ indicates the onset of a new dielectric mode which turns on below $32K$. We describe this mode with parameters $\varepsilon_{\alpha 0}(T) = 900(1 - T/T_{d\alpha})$ with $T_{d\alpha} = 32K$ and $\tau_\alpha(T) = 6.5 \times 10^{-9}(\sec \cdot K)/T$. $\varepsilon_{\alpha 0}(T)$ is similar to an order parameter which grows below a transition. As T is lowered, $\tilde{\varepsilon}_\alpha(T)$ increases initially due to gradual displacement of oxygen leading to growing polarization, and is well described by a model based upon octahedral distortions [14]. However below a characteristic temperature $\tilde{\varepsilon}_\alpha(T)$ begins to decrease because the dipoles are no longer able to follow the microwave field.

(2) In $La_{5/3}Sr_{1/3}NiO_4$ a transition is observed at exactly the same temperatures $32K$ as in $La_2CuO_{4.0175}$. However the $32K$ dielectric mode in $La_{5/3}Sr_{1/3}NiO_4$ has a much weaker strength with $\varepsilon_{\alpha 0}(T) \sim 0.5(1 - T/T_{d\alpha})$, and requires a high sensitivity measurement such as the present microwave measurement. $\tilde{\varepsilon}_\beta(T)$ of $La_2CuO_{4.0175}$ associated with a $T_{d\beta} = 245K$ transition is described by $\varepsilon_{\beta 0}(T) = 10(1 - T/T_{d\beta})$ with $\tau_\beta(T) = 1 \times 10^{-9}(\sec \cdot K)/T$. $\tilde{\varepsilon}_\gamma$(T) which is dominant at higher T is described with a relaxation

time $\tau_\gamma(T) = 8 \times 10^{-13}\,\mathrm{sec}^{-1}\exp(1000/T)$ characterized by an activation energy $1000K$. Since $La_{5/3}Sr_{1/3}NiO_4$ and $La_2CuO_{4.0175}$ are isostructural the identical temperature scales suggest the same structural origin, *viz.* octahedral instabilities, as discussed in ref.[14]

(3) In insulating $YBa_2Cu_3O_{6.0}$, the parent compound of the 123 family, microwave dielectric transitions are observed at $60K$ and $110K$ [12]. These transitions were identified with lattice instabilities - e.g. change in buckling angle for the $110K$ transition. Similar polar structures were observed also in the measurements of pyro- and piezoelectricity[13] of $YBa_2Cu_3O_{7-\delta}$.

These results clearly show that there are important temperature scales : $32K$ and $245K$ in the $La-214$ compounds, and $60K$ and $110K$ in the $Y-123$ compounds, that we identify with local lattice instabilities. Indeed a careful examination of the literature shows that anomalies at these temperatures are observed in several experiments (see references in [14]). Additional temperature scales may well be present, including at very high temperatures $> 300K$. The presence of these temperature scales underscores the importance of the lattice, which has largely been ignored.

As noted earlier, the microwave response is entirely due to the charge system, which of course is in turn coupled to the lattice and the spins. Thus lattice instabilities affect the charge system and hence the microwave response. While we have shown this clearly in the parent insulating state, these signatures persist even into the doped conducting or even superconducting state.

5. Coherent explanation of the electrodynamic response of the cuprates using a collective mode approach

As discussed above the collective mode description of the microwave conductivity provides an excellent description of the electrodynamic response of the cuprates in a variety of doping and temperature scales. This is summarized in Fig.1.

(1) In the parent insulating compound the response is that of dielectric modes, which can be viewed as the strongly pinned limit $\omega \ll \omega_{pin}$ of the conductivity collective mode. Lattice instabilities lead to onset of these modes at specific temperatures. Some of these temperatures coincide with stripe formation temperatures.

(2) For moderate doping and high temperatures in the pseudogap state,

a microwave plasmon, characterized by a very narrow conductivity mode with width $\Gamma \sim 0.1meV$ is required for explaining the anomalous features observed in the microwave response. Here the Drude limit of the pinned mode applies, i.e. $\omega_{pin} \to 0$. There is evidence for the narrow collective mode even well above $T_6 < c$ at T as high as $300K$. A microscopic explanation of the available data is feasible in a scenario of fluctuating stripes.

(3) The microwave response in the superconducting state is completely dominated by the collective mode response in the Drude limit. Here too $\Gamma < 0.1meV$. In addition when $\omega \sim \Gamma(T)$, a conductivity peak occurs. The relative differences among the cuprates are entirely due to varying Γ. This collective mode conductivity contribution dominates the superconducting quasiparticle contribution. In contrast to the quasiparticle scenario, the oscillator strength of the conductivity mode increases with decreasing temperature.

(4) Multiple conductivity peaks are commonly observed, for example in $YBa_2Cu_3O_{6.5}$ in Fig.5, and may indicate the presence of multiple conductivity modes. These modes have different scattering rates $\tau_i(T)$ and hence different peak temperatures given by $\omega\tau_i(T_{ip}) = 1$. The multiple conductivity mode scenario in the doped conducting state is similar to the multiple dielectric modes observed in the insulating states. An intriguing possibility exists that the conductivity (dielectric) modes represent the unpinned (pinned)limit of the same collective mode. A more precise understanding of the evolution with doping of the dielectric and conductivity modes remains to be established.

(5) It should be noted that although we have used the CDW picture, it is possible that the conductivity mode discussed above originates from current fluctuations rather than charge fluctuations. Further experiments are needed to distinguish between these possibilities.

The above work has shown unambiguously the presence of a narrow collective mode. FIR measurements by Tajima [15], et.al., are also consistent with the above conclusions.

Charge inhomogeneities (e.g.stripes) as microscopic origin for the conductivity mode

A microscopic origin for the macroscopic collective conductivity mode discussed above is certainly required. Towards this end,an examination of the evolution of the charge and spin systems with doping is illuminating.

Spin probes, such as NS, muSR and NMR, have shown that a spin glass state persists upon doping and even into the superconducting state. The presence of magnetic stripes at almost all doping is now well established. The present work shows that a similar situation occurs for the charge system also. Long-lived charge (and/or current) fluctuations characterized by time scales $\sim 10^{-12}$ sec are present deep into the superconducting state. The dimensionality of these fluctuations, whether stripes or blobs, is not clear from the present macroscopic measurements, but their time scale, and the narrowness of the accompanying collective mode, is unquestionable.

We thank numerous collaborators for providing high quality single crystal samples discussed in this work. This work was supported by the Office of Naval Research.

References

1. G. Gruner, "Density Waves in Solids", Addison-Wesley, Reading,MA (1994).
2. For further details on samples, see the following. A. Erb, et. al., J. Phys. Chem. Solids **59**, 2180 (1998); A.Sacuto, et al. Phys. Rev. B **58**, 11721-11733 (1998); J. R. Kirtley, et al., Phys. Rev. Lett. **81**, 2140-2143 (1998); S. - W. Cheong et al. Phys. Rev. Lett. **79**, 2514, (1997); R. J. Gooding, et al. Phys. Rev. B **55**, 6360-6371 (1997).
3. Z.Zhai, et al., Rev. Sci. Instr. **71**, 3151 (2000).
4. S. Sridhar and Z. Zhai, Physica C : Proc. of M2S-HTSC-VI, **341-348**, 2057 (2000)
5. D. C. Mattis and J. Bardeen, Phys. Rev. **111**, 412 (1958).
6. T. Jacobs, et al., Phys. Rev. Lett. **75**, 4516 (1995).
7. J. R. Waldram, et. al., Phys. Rev. B **59** 1528 (1999)
8. D. A. Bonn, et. al., Phys. Rev. Lett. **68**, 2390 (1992).
9. S. Chakravarty, R.B. Laughlin, C. Nayak, Phys. Rev. B **63**, 094503 (2001)
10. N. Hasselmann, et.al., Phys. Rev. Lett. **82**, 2135 (1999).
11. C.Kusko, et. al., Phys. Rev. B., **65**, 132501 (2002)
12. Z. Zhai, et al., Phys. Rev. B **63**, 092508 (2001)
13. J. Demsar, et. al., Phys. Rev. Lett., **82**, 4918 (1999).
14. P. V. Parimi, N. Hakim, F. C. Chou, S. W. Cheong and S. Sridhar, LANL archives cond-mat 0108342.
15. S. Tajima, International Conference on the Low Energy Electrodynamics in Solids LEES '02, Montauk, NY, (unpublished)

NONLINEAR ELASTICITY, MICROSTRUCTURE AND COMPLEX MATERIALS

A. SAXENA, T. LOOKMAN AND A. R. BISHOP

Theoretical Division,
Los Alamos National Lab,
Los Alamos, NM 87544, USA
E-mail: avadh@lanl.gov

S. R. SHENOY

Abdus Salam International Centre for Theoretical Physics
Trieste 34100, Italy
E-mail: shenoy@ictp.trieste.it

We review intrinsic spatial variations of the order parameter (e.g. strain, polarization, magnetization, etc.) in a variety of materials undergoing transitions such as ferroelastic martensites, shape memory alloys, manganites, cuprates, relaxor ferroelectric titanates, and magnetoelastics. The usefulness of strain tensor and elastic compatibility constraint in providing key physical insights as well as a full description (including microstructure) of materials undergoing structural transitions is emphasized. We outline applications of these ideas, in conjunction with recent experiments, for understanding strain textures such as twinning and transformation precursors, i.e. modulated phases.

1. Introduction

Complex functional materials often involve structural phase transitions between different crystalline phases. The latter have attracted a great deal of interest for over a century, both for their conceptual importance as symmetry-changing phase transitions, and for their role in inducing technologically useful materials properties. Both the diffusion-controlled replacive and the diffusionless displacive transformations have been studied, although the former have received more attention because their reaction kinetics is more conducive to control.

We consider here the class of materials known as ferroelastics. Ferroelasticity is defined by the existence of two or more stable orientation states of a crystal that correspond to different arrangements of the atoms, but are structurally identical[1,2]. In addition, these orientation states are degenerate in energy in the absence of mechanical stress. Salient features of ferroelastic crystals include mechanical hysteresis and mechanically (reversibly) switchable domain patterns.

Usually ferroelasticity occurs as a result of a phase transition from a non-ferroelastic high-symmetry 'parent' phase and is associated with the softening of an elastic modulus with decreasing temperature or increasing pressure in the parent phase. Since the ferroelastic transition is normally weakly first order, or second order, it can be described to a good approximation by the Landau theory with spontaneous strain or deviation of a given ferroelastic orientation state from the parent phase as the order parameter. The strain can be coupled to intra-unit cell (shuffle) modes or other fields such as electric polarization and magnetic moment, and thus the crystal can have more than one transition. Depending on whether the spontaneous strain is the primary or a secondary order parameter at a given transition, the lower symmetry phase is called a proper or an improper ferroelastic, respectively. While martensites (materials undergoing first order, diffusionless transition[3]) are proper ferroelastics, examples of improper ferroelastics include ferroelectrics[4], magnetoelastics[5], colossal magnetoresistance[6] (CMR) and high-temperature superconducting materials[6].

There is a further subset of ferroelastic martensites (either nonelemental metals or alloy systems) that exhibit the shape memory effect[7]. These materials are characterized by highly mobile twin boundaries and (often) show precursor structures (e.g., tweed[8,9,10]) above the transition. Furthermore, these materials have small Bain strain, elastic shear modulus softening, and a weakly to moderately first order transition. Some examples include InTl, FePd, NiTi and AuCd. To optimize the desirable functionalities of these materials it is crucial to understand microstucture such as twinning and tweed. Recently, magnetic shape memory effect has been observed in high temperature superconductors[11]. In addition to martensites and shape memory alloys, tweed and related modulated phases have been observed in other materials[8] such as quartz, high T_c superconducting perovskites[9,12,13], GMR materials[14], CMR materials[13], ferroelectrics[15], magnetic martensites[16], etc. Both polar[15] and magnetic modulated phases[17,18] have been observed.

Recent advances[6] in high resolution (scanning and transmission) electron microscopies, local structure and microstructure characterization probes (pair distribution function analysis, neutron scattering with high-intensity pulsed sources, x-ray scattering with advanced photon sources, resonant ultrasound, etc.) as well as concurrent developments in new ways of modeling complex electronic/elastic materials are indeed beginning to enable a deeper understanding of multiscale properties of these materials, which can lead to control over various key functionalities. Our focus here is on materials where elasticity plays a crucial role [e.g. as an order parameter (OP) in a solid-solid phase transition]. Our approach to the inherent complexity and collective, multiscale, phenomena is to merge techniques from traditional materials science, statistical physics and nonlinear condensed matter to relate materials texture to local (e.g. electronic or magnetic) functionalities.

Examples include (i) magnetoelastics[5] (e.g. Terfenol: $Tb_xDy_{1-x}Fe_2$ which has the highest known magnetostrictive coefficient) with magnetization as the primary OP, (ii) piezoelectrics, ferroelectrics and antiferroelectrics ($BaTiO_3$, $SrTiO_3$, PZT which has among the highest known electrostrictive coefficients[19]) usually with polarization as the primary OP, (iii) CMR perovskites[6] ($La_xSr_{1-x}MnO_3$), high temperature superconductors[6] (La_xCuO_4, $YBaCuO_7$), and photoelastics[20] ($GaIn_xAs_{1-x}$) with optical birefringence as the primary property. Certain materials of geological[21] and planetary[22] interest (iron, silicates), biological interest (e.g. proteins in bacteriophage viruses[23]), actinides[24] and with self-similar triangular twinning[25] [$Pb_3(VO_4)_2$] also belong to the above class.

In all of the above complex electronic/elastic materials, along with the atomic scale inhomogeneities of charge, spin and lattice distortion, have come observations of "fine-scale structure" on more materials science scales[6] (100s of lattice constants and beyond). These observations of twinning, tweed, etc., appear in extensive precursor regimes of global solid-solid phase transformations or as functions of mechanical stress. This multiscale structure is increasingly understood to produce anomalous elastic constants[26], low-symmetry lattice vibrations, low-frequency scattering in x-ray and neutron spectra, and inhomogeneous coupling to electronic, magnetic, ferroelectric and superconducting degrees of freedom. The associated multiscale dynamics is intimately involved in shape-memory[7], relaxor[27] and related phenomena.

The mechanism by which local perturbations (such as localized charge patterns) can have large-scale effects on elastic patterning is *anisotropic long-range strain-strain coupling.* This *long-range* coupling arises because of elastic compatibility constraints on the allowed *unit cell* deformations (in the absence of bond-breaking) and is therefore indeed inherited from very *local*, atomic scale, symmetries. We have recently developed models of elasticity and solid-solid phase transformations at the coarse-graining level of a nonlinear Ginzburg-Landau description, including the compatibility constraint self-consistently. This allows a description of systematic twinning, tweed and other fine-scale hierarchical structures, *the connection of intrinsic multiscale phenomena in solid state physics and materials science.* Our model can predict multiscale elastic textures and their consequences in a unified manner. A long-range, anisotropic strain field can be produced by the local stresses of doped charges in transition metal oxides, or other materials with significant elastic coupling.

2. Elastic Compatibility

The true degrees of freedom in a continuum elastic medium (such as described by Ginzburg-Landau models) are contained in the displacement field, $u(r)$, even though it is the strain fields, ϵ_{ij}, which appear in the free energy. Instead of treating the strains as independent fields, one must assume that they correspond to a physical displacement field, i.e., that they are derivatives of a single continuous function. This is achieved by requiring that they satisfy a set of nontrivial compatibility relations concisely expressed as a differential relation between the strain components. Ignoring the geometrical compatibility constraint and minimizing the free energy directly would lead to the incorrect result that non-OP strain components are identically zero, and the OP strain trivially responds to perturbations, e.g. stress and local disorder, which is certainly not true. One can explicitly account for the compatibility constraint by appending it to the free energy via a Lagrange multiplier. Then the non-OP strain components can be expressed in terms of OP strain components by minimizing the free energy. This procedure results in a nonlocal (anisotropic, long-range) interaction between the OP strain $e(r)$ and $e(r')$.

In d-dimensions the displacement field has d independent components at any point $\mathbf{x}$, whereas a symmetric tensor nominally has $d(d+1)/2$ independent components. Because the strain tensor is composed of derivatives of a vector field, there must be relations or constraints among its components, so that all components can not vary in arbitrary ways. In the approximation of "geometrical linearity" these constraints are expressed by the Saint-Venant compatibility relation[28,29,30].

The components of the Lagrangian strain tensor in any dimension are defined by

$$\epsilon_{ij} = \frac{1}{2}(u_{i,j} + u_{j,i} + u_{k,i}u_{k,j}), \tag{1}$$

where $u_{i,j} = \partial u_i/\partial x_j$ are the derivatives of the displacement vector with respect to material coordinates x_j in a Cartesian frame. The last term in the above equation (with summation over k implied) refers to 'geometrical nonlinearity" in strain and is important for finite strain deformations and lattice rotation. However, in most structural transitions of interest the strain is ususally less than 10 % and the last term can be neglected. In this approximation of geometrically linear elasticity, the compatibility condition (in any dimension) is compactly written as[28,29,30]

$$\nabla \times (\nabla \times \overleftrightarrow{\epsilon})^T = 0, \tag{2}$$

where T denotes transpose. The zero on the right hand side indicates no source terms such as arising from dislocations or disclinations. The "incompatibility" due to such lattice topological defects can be included by using the Burgers vector (density) on the right hand side of the equation. In two dimensions (2D) there is only one compatibility equation:

$$\epsilon_{xx,yy} + \epsilon_{yy,xx} = 2\epsilon_{xy,xy}. \tag{3}$$

However, in three dimensions (3D) there are six compatibility equations (in two sets of three equations each):

$$\begin{aligned}
\epsilon_{yy,zz} + \epsilon_{zz,yy} &= 2\epsilon_{yz,yz}, \\
\epsilon_{xx,zz} + \epsilon_{zz,xx} &= 2\epsilon_{xz,xz}, \\
\epsilon_{xx,yy} + \epsilon_{yy,xx} &= 2\epsilon_{xy,xy}, \\
\epsilon_{xx,yz} + \epsilon_{yz,xx} &= \epsilon_{xy,xz} + \epsilon_{xz,xy}, \\
\epsilon_{yy,xz} + \epsilon_{xz,yy} &= \epsilon_{xy,yz} + \epsilon_{yz,xy}, \\
\epsilon_{zz,xy} + \epsilon_{xy,zz} &= \epsilon_{xz,yz} + \epsilon_{yz,xz}.
\end{aligned}$$

Note that the 2D compatibility equation has a symmetry under coordinate transformations such that the term x^2y^2 remains invariant. Similarly, the 3D compatibility equations have a symmetry such that the terms x^2y^2 and x^2yz (and their two appropriate permutations $x \to y \to z \to x$) are invariant under coordinate transformations.

3. Ferroelastic transitions

In order to identify the components of the strain tensor that serve as the primary OP in a structural transition, symmetry methods are very convenient, particularly in 3D. Strain is a macroscopic variable and corresponds to $k = 0$, i.e. the Brillouin zone center or the Γ point. Thus the (reducible representation for the) strain tensor is labeled by Γ_ϵ which can be decomposed into irreducible representations (IR) according to the crystal symmetry as demonstrated in Table I for 2D and Table II for 3D. Since strain changes only the unit cell shape (but not its size, i.e. no cell doubling, tripling, etc.) the strain free energy does not depend on lattice translations and is thus invariant under the point group of the crystal. The one-dimensional IR Γ_1 (in 2D with the basis function of the form $x^2 + y^2$) or Γ_1^+ (in 3D with the basis function of the form $x^2 + y^2 + z^2$) corresponds to the dilatation of the unit cell without any change in symmetry. Thus the ferroelastic transitions are driven by other IRs corresponding to either deviatoric or shear strains. A particular IR (Γ_1^+ in general and specifically Γ_3^+ for the trigonal case and Γ_2^+ for the monoclinic case) can appear more than once indicating that different strain tensor components can serve as basis functions for the *same* IR. There are 23 proper ferroelastic transitions[31] in 2D and 94 proper ferroelastic transitions[32] in 3D.

Table I. Proper ferroelastic transitions and strain irreducible representations (IR) for the four crystal systems in 2D.

Lattice	Group (No.)	Strain IR
Hexagonal	p6mm (17)	$\Gamma_\epsilon = \Gamma_1 \oplus \Gamma_5$
Square	p4mm (11)	$\Gamma_\epsilon = \Gamma_1 \oplus \Gamma_2 \oplus \Gamma_3$
Rectangular	p2mm (6)	$\Gamma_\epsilon = 2\Gamma_1 \oplus \Gamma_3$
Oblique	p2 (2)	$\Gamma_\epsilon = 2\Gamma_1 \oplus \Gamma_2$

Table II. Proper ferroelastic transitions and strain irreducible representations (IR) for the seven crystal systems in 3D.

Lattice	Group (No.)	Strain IR
Cubic	Pm$\bar{3}$m (221)	$\Gamma_\epsilon = \Gamma_1^+ \oplus \Gamma_3^+ \oplus \Gamma_5^+$
Tetragonal	P4/mmm (123)	$\Gamma_\epsilon = 2\Gamma_1^+ \oplus \Gamma_2^+ \oplus \Gamma_4^+ \oplus \Gamma_5^+$
Orthorhombic	Pmmm (47)	$\Gamma_\epsilon = 3\Gamma_1^+ \oplus \Gamma_2^+ \oplus \Gamma_3^+ \oplus \Gamma_4^+$
Hexagonal	P6/mmm (191)	$\Gamma_\epsilon = 2\Gamma_1^+ \oplus \Gamma_5^+ \oplus \Gamma_6^+$
Trigonal	P$\bar{3}$1m (162)	$\Gamma_\epsilon = 2\Gamma_1^+ \oplus 2\Gamma_3^+$
Monoclinic	P2/m (10)	$\Gamma_\epsilon = 4\Gamma_1^+ \oplus 2\Gamma_2^+$
Triclinic	P$\bar{1}$ (2)	$\Gamma_\epsilon = 6\Gamma_1^+$

For improper ferroelastic transitions such as in ferroelecrtics, cuprates and manganites the nonlinear free energy is expanded in the relevant primary OP (e.g. polarization, magnetization, shuffle, octahedral tilt angle, charge density, etc.) consistent with the crystal symmetry (and the IRs of the primary OP) and the strain is included only to the harmonic order. Similarly, symmetry allowed couplings between the primary OP and strain are included to the lowest order (see Section 4.2).

3.1. *2D transitions*

In 2D (see Table I) we have previously studied the p4mm (C_{4v}^1: 2D space group 11) to p2mm (C_{2v}^1: 2D space group 6) ferroelastic transition which takes a square lattice to a rectangular lattice with two orientational states of the latter[33]. For the sake of illustration, here we consider a ferroelastic transition which takes a triangular lattice p6mm (C_{6v}^1: 2D space group 17) to a centered rectangular lattice c2mm (2D space group 9) with three orientational states of the latter. The OP strain is a two-component shear $\Gamma_5 = (\epsilon_2, -\epsilon_3)$ and the non-OP strain is the dilatation $\Gamma_1 = (\epsilon_1)$. We use the symmetry adapted forms for strain $\epsilon_1 = \frac{1}{2}(u_{x,x} + u_{y,y})$, $\epsilon_2 = \frac{1}{2}(u_{x,x} - u_{y,y})$ and $\epsilon_3 = \frac{1}{2}(u_{x,y} + u_{y,x})$. The associated local rotation is given by $\omega_3 = \frac{1}{2}(u_{x,y} - u_{y,x})$.

The free energy consists of three parts $F = F_L + F_C + F_G$. The Landau term without coupling (F_L), the part with coupling between the OP and non-OP strains (F_C), and the Ginzburg part containing gradients of the OP strain (F_G). The free energy for this transition, to fourth degree in OP and to second degree in non-OP strain ϵ_1, is given in Refs. [31,34]:

$$F_L = A_1\epsilon_1^2/2 + A_2(\epsilon_2^2 + \epsilon_3^2)/2 + B(\epsilon_3^3 - 3\epsilon_2^2\epsilon_3)/3 + C(\epsilon_2^2 + \epsilon_3^2)^2/4 \tag{4}$$

$$F_C = D\epsilon_1(\epsilon_2^2 + \epsilon_3^2) \tag{5}$$

$$\begin{aligned} F_G = {} & g_1(3\epsilon_{2,x}^2 + 2\epsilon_{2,x}\epsilon_{3,y} + \epsilon_{2,y}^2 + 2\epsilon_{2,y}\epsilon_{3,x} + \epsilon_{3,x}^2 + 3\epsilon_{3,y}^2) \\ & + g_2(\epsilon_{2,x}^2 + 6\epsilon_{2,x}\epsilon_{3,y} - \epsilon_{2,y}^2 - 2\epsilon_{2,y}\epsilon_{3,x} - \epsilon_{3,x}^2 + \epsilon_{3,y}^2) \\ & + g_3(\epsilon_{2,x}^2 - 2\epsilon_{2,x}\epsilon_{3,y} + 3\epsilon_{2,y}^2 - 2\epsilon_{2,y}\epsilon_{3,x} + 3\epsilon_{3,x}^2 + \epsilon_{3,y}^2). \end{aligned} \tag{6}$$

Here A_1 and A_2 are the bulk and shear modulus, respectively. B and C are the third and fourth order elastic constants, and g_i denote strain gradient coefficients which can be obtained from the curvature of the phonon dispersion curves in the vicinity of the Brillouin zone center (Γ point).

Since a third-order invariant is allowed in F_L, this transition is of first order. It is straightforward to include sixth degree terms in the OP for those first order transitions in which a third-order invariant is not symmetry allowed[35,36]. The compatibility equation (in terms of symmetry adapted strains) for all transitions in 2D is of the form[33,37]

$$G \equiv (\partial_x^2 + \partial_y^2)\epsilon_1 - (\partial_x^2 - \partial_y^2)\epsilon_2 - 2\partial_x\partial_y\epsilon_3 = 0. \tag{7}$$

The Landau term can be further written as $F_L = F_0 + F_{comp}$.

The Euler-Lagrange variation of $[F - \Lambda G]$ with respect to the non-OP strains, is then[33] $\delta(F_{comp} - \Lambda G)/\delta\epsilon_1 = 0$. In the case of a triangular lattice $F_{comp} = \sum A_1\epsilon_1^2/2$ is identically equal to $\sum_k F_{comp}(k)$. The variation gives (in k space assuming periodic boundary conditions)

$$\epsilon_1(k) = (k_x^2 + k_y^2)\Lambda(k)/A_1. \tag{8}$$

We then put $\epsilon_1(k)$ back into the compatibility constraint condition and solve for the Lagrange multiplier $\Lambda(k)$. Thus $\epsilon_1(k)$ is expressed in terms of $\epsilon_2(k)$, $\epsilon_3(k)$ and

$$F_{comp}(\vec{k}) = A_1|(k_x^2 - k_y^2)\epsilon_2/k^2 + 2k_x^2k_y^2\epsilon_3/k^2|^2/2, \tag{9}$$

identically equal to $(1/2)A_1U_{\ell\ell'}(\vec{k})\epsilon_\ell(\vec{k})\epsilon_{\ell'}(\vec{k})$ with $l = 2, 3$, which is used in a (static) free energy variation of the OP. The (static) "compatibility kernel" $U(\hat{k})$ is independent of $|\vec{k}|$ at long-wavelengths: $U(\vec{k}) \to U(\hat{k})$. In coordinate space this is an anisotropic long-range ($\sim 1/r^2$) potential mediating the elastic interactions of the primary OP strain.

The Ginzburg-Landau free energies and compatibility kernels for the ferroelastic transitions in the four 2D crystal systems are listed in Ref. [31].

3.2. *3D transitions*

For the cubic system (space group 221: Pm$\bar{3}$m, O_h^1) we define the following symmetry-adapted (unnormalized) strains: the dilatation $e_1 = \epsilon_{xx} + \epsilon_{yy} + \epsilon_{zz}$ which is the basis function for the IR Γ_1^+, two deviatoric strains $e_2 = \epsilon_{xx} + \epsilon_{yy} - 2\epsilon_{zz}$ and $e_3 = \epsilon_{xx} - \epsilon_{yy}$ which serve as the basis functions for the IR Γ_3^+, and three shear strains $e_4 = \epsilon_{xy}$, $e_5 = \epsilon_{yz}$ and $e_6 = \epsilon_{xz}$ as the basis functions for the IR Γ_5^+.

Cubic$\to$Tetragonal (or orthorhombic): This transition is driven by the two-dimensional IR Γ_3^+ with basis vectors which are the two deviatoric order parameter (OP) strains (e_2, e_3). The free energy $F = F_{OP} + F_{NOP}$ is given by[38]

$$F_{OP}(e_2, e_3) = a(e_2^2 + e_3^2) + b(e_2^3 - 3e_2 e_3^2) + c(e_2^2 + e_3^2)^2,$$

$$F_{NOP}(e_1, e_4, e_5, e_6) = \frac{A_1}{2} e_1^2 + \frac{A_4}{4}(e_4^2 + e_5^2 + e_6^2)$$

$$= \int d^3r e_2(r) U_{22}(r, r') e_2(r') + \int d^3r e_3(r) U_{33}(r, r') e_3(r')$$

$$+ \int d^3r e_2(r) U_{23}(r, r') e_3(r'),$$

where $U_{ij}(r, r')$ are anisotropic, long-range (ALR) elastic compatibility kernels obtained by minimizing the non-OP strains using the six compatibility equations as constraints. When (e_2, e_3)= (e_0, 0) a tetragonal structure is obtained and when (e_2, e_3) = (e_a, e_b) an orthorhombic phase is obtained. We have studied the elastic compatibility induced anisotropic long-range interaction and microstructure for this transition previously[38]. The (two invariants of the) associated strain gradient energy F_G can be found there.

Cubic→Trigonal (or orthorhombic or monoclinic or triclinic): This transition is driven by the three-dimensional IR Γ_5^+ with basis vectors which are the three shear OP strains (e_i: e_4, e_5, e_6). The free energy $F = F_{OP} + F_{NOP}$ is given by

$$F_{OP}(e_i) = a(e_4^2 + e_5^2 + e_6^2) + b e_4 e_5 e_6 + c(e_4^2 + e_5^2 + e_6^2)^2 + d(e_4^4 + e_5^4 + e_6^4),$$

$$F_{NOP}(e_1, e_2, e_3) = \frac{A_1}{2} e_1^2 + \frac{A}{2}(e_2^2 + e_3^2)$$

$$= \int d^3r e_4(r) U_{44}(r, r') e_4(r') + \int d^3r e_5(r) U_{55}(r, r') e_5(r')$$

$$+ \int d^3r e_6(r) U_{66}(r, r') e_6(r') + \int d^3r e_4(r) U_{45}(r, r') e_5(r')$$

$$+ \int d^3r e_4(r) U_{46}(r, r') e_6(r') + \int d^3r e_5(r) U_{56}(r, r') e_6(r').$$

When (e_4, e_5, e_6)=(e_0, 0, 0) an orthorhombic phase is obtained, when (e_4, e_5, e_6)=(e_0, e_0, e_0) a trigonal phase is obtained, when (e_4, e_5, e_6)=(e_a, e_a, e_b) a monoclinic phase is obtained, and when (e_4, e_5, e_6)=(e_a, e_b, e_c) a triclinic phase is obtained. Note that both of the above free energies for a cubic crystal correspond to a first order phase transiton since a third order invariant is symmetry allowed. The (four invariants) of the gradient energy F_G for this transition are:

$$F_G = g_1(e_{4x}^2 + e_{4y}^2 + e_{5y}^2 + e_{5z}^2 + e_{6x}^2 + e_{6z}^2) + g_2(e_{4x} e_{5z} + e_{4y} e_{6z} + e_{5y} e_{6x})$$

$$+g_3(e_{4z}^2 + e_{5x}^2 + e_{6y}^2) + g_4(e_{4z}e_{5x} + e_{4z}e_{6y} + e_{5x}e_{6y}).$$

The Tetragonal→Orthorhombic transition occurs in cuprates, manganites and many other functional materials. From Table II we note that this transition can be driven by either of the two one-dimensional IRs Γ_2^+ (with strain e_3) or Γ_4^+ (with strain e_4). For the IR Γ_2^+, the Ginzburg-Landau free energy is given by:

$$F_{OP}(e_3) = ae_3^2 + be_3^4 + ce_3^6 + g(\nabla e_3)^2,$$

$$F_{NOP}(e_1, e_2, e_4, e_5, e_6) = \frac{A_1}{2}e_1^2 + \frac{A_2}{2}e_2^2 + \frac{A_4}{2}e_4^2 + \frac{A_5}{2}(e_5^2 + e_6^2)$$

$$= \int d^3 r e_3(r) U_{33}(r, r') e_3(r').$$

The Ginzburg-Landau free energies for ferroelastic transitions arising from other non-cubic crystal systems can similarly be obtained using Table II and Ref. [36].

4. Microstructure in Complex Materials

There are three sources of anisotropy in the microstructure, namely from the Landau part of the free energy (only for cases involving multi-component order parameters), the Ginzburg part (only in cases where gradient invariants other than $g(\nabla e)^2$ are allowed) and the compatibility-induced long-range interaction (which is always anisotropic).

4.1. *Proper ferroelastics*

Microstructure simulation for the Square→Rectangle transition based on a relaxational dynamics, namely the time-dependent Ginzburg-Landau (TDGL) dynamics, are given in Ref. [33]. A typical twinning (below the martensitic transition temperature) and tweed microstructure (above the transition temperature) are depicted there. The latter shows strain striations along the ±45° directions.

We have recently derived a proper ferroelastic strain dynamics including damping and noise[34] which reduces to the TDGL dynamics under appropriate conditions. Simulations based on this underdamped dynamics are given in Ref. [34], where the evolution of twins between the two rectangular variants can be clearly seen. Note that this simulation serves as a 2D analog of the Cubic→Tetragonal transition in the shape memory alloy FePd (the full 3D simulations of twinning for this transition are presented in Ref.

[38]) and the Tetragonal→Orthorhombic transition in high tempererature superconductor[39] $YBa_2Cu_3O_{7-\delta}$.

A Triangle→Rectangle transition microstructure evolution is depicted in Fig. 1 starting from random initial conditions using the TDGL dynamics. A nontrivial microstructure, i.e. nested stars involving the three rectangular orientations, emerges. Note that this microstructure has been observed experimentally[25]. This simulation serves as a 2D analog of the Hexagonal→Orthorhombic transition in $Pb_3(VO_4)_2$ and the Trigonal→Orthorhombic transition in CMR perovskite material[40] $La_{1-x}Sr_xMnO_3$.

4.2. *Improper ferroelastics*

As a representative improper ferroelastic transition, we consider a 2D ferroelectric on a square lattice with polarization $\mathbf{P}=(P_x, P_y)$. We first define symmetry-adapted polarizations (akin to symmetry-adapted strains):

$$R = P_x^2 + P_y^2, \quad S = P_x^2 - P_y^2, \quad T = P_x P_y. \tag{10}$$

The total free energy is written as $F = F_P + F_e + F_{eP}$, where

$$\begin{aligned} F_P &= \alpha_1 R + \alpha_2 R^2 + \alpha_3 S^2 + \alpha_4 T^2 + ... + \frac{g}{2}(\nabla P)^2, \\ F_e &= \frac{A_1}{2}\epsilon_1^2 + \frac{A_2}{2}\epsilon_2^2 + \frac{A_3}{2}\epsilon_3^2, \\ F_{eP} &= \beta_1 \epsilon_1 R + \beta_2 \epsilon_2 S + \beta_3 \epsilon_3 T. \end{aligned}$$

Using elastic compatibility constraint and minimizing the total energy with respect to strain tensor components we find that the elastic energy can be expressed in terms of an anisotropic long-range interaction between (second moments of) symmetry-adapted polarizations:

$$\begin{aligned} F_e = &\sum_r R(r)X(|r-r'|)R(r') + \sum_r S(r)Y(|r-r'|)S(r') \\ &+ \sum_r T(r)Z(|r-r'|)T(r'). \end{aligned}$$

Here $X(|r-r'|)$, $Y(|r-r'|)$ and $Z(|r-r'|)$ are long-range kernels which are known exactly in the Fourier (or k) space. The above interaction can be added to F_P to simulate microstructure in a 2D ferroelectric or analogously a 2D magnetoelastic material with square symmetry. Similarly, symmetry allowed strain couplings to charge (doping), magnetization as well as intra-unit cell distortions (or shuffle modes) in improper ferroelastics such

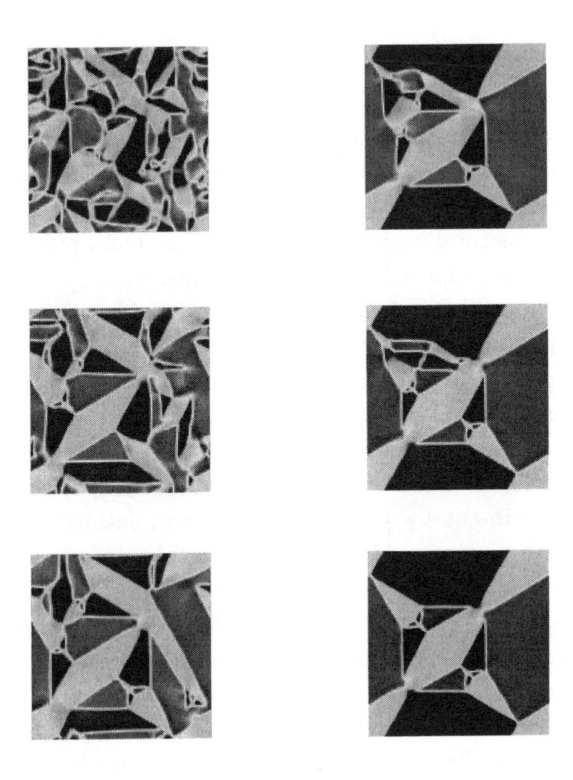

Figure 1. Various snapshots of microstructure evolution from random initial conditions for the triangular lattice to (centered) rectangular lattice transition. The three shades of gray represent the three rectangular orientations. The emergence of a nested star pattern is clearly seen.

as manganites and cuprates can explain intrinsically inhomogeneous spin, charge and lattice textures[41] (e.g. orbital ordering, stripe segments).

5. Conclusion

The importance of ferroelastic strain in CMR materials[42] and in high temperature superconductors within the context of nanodomain structures[43,44] is becoming increasingly clear. We provided a systematic approach to strain modeling when a structural transition is also present. Our approach is in the spirit of the Landau description of phase transitions: working with the order parameters as the basic and physically relevant variables, and focusing on the order parameter symmetries (as encoded in the compatibility factors), as the source of proper/improper ferroelastic static (and dynamic) texturing. A natural extension of our work includes a detailed understanding of $2D$ nucleation, growth and interfacial profiles; generalizations to include defects, in a broader 'strain elastodynamics' framework; simulations and calculations of experimentally measurable strain correlations and nonlinear susceptibilities. We have recently developed a strain dynamics[34] properly taking into account the elastic compatibility constraint.
This symmetry-specific, compatibility-focused ferroelastic statics/dynamics for the strain order parameter encode, in their very structure, the possibility of an evolutionary textural hierarchy in both space and time, and a tendency for interfaces to be driven at sound speeds, explaining some of the fascinating but puzzling features of martensitic dynamics. The dynamical equations[34] can now be applied to a wide variety of textural evolutions that include improper ferroelastics (in particular manganites and cuprates), leading to a deeper understanding of many materials of technological interest such as ferroelectrics, magnetoelastics, colossal magneto-resistance manganites, superconducting cuprates, and shape memory materials. Our approach can be readily applied to find the orientations of domain walls in the primary OP, e.g. ferroelectric and magnetoelastic walls or interfaces between charge- and orbital-ordered regions.

6. Acknowledgment

We thank R. C. Albers, Yu. B. Gaididei, D. M. Hatch and K. Ø. Rasmussen for fruitful discussions. This work was supported by the U.S. Department of Energy.

References

1. V. K. Wadhawan, *Introduction to Ferroic Materials* (Gordon and Breach, Amsterdam, 2000).
2. E.K.H. Salje, *Phase Transformations in Ferroelastic and Co-elastic Solids* (Cambridge University Press, Cambridge, U.K., 1990).
3. Z. Nishiyama, *Martensitic Transformations* (Academic, New York, 1978).
4. M. E. Lines and A. M. Glass, *Principles and Applications of Ferroelectrics and Related Materials* (Clarendon Press, Oxford, U.K., 1977).
5. R. D. James and M. Wuttig, Phil. Mag. A **77**, 1273 (1998).
6. See various contributions in this volume.
7. *Shape Memory Materials*, edited by K. Otsuka and C. M. Wayman (Cambridge University Press, Cambridge, U.K., 1998); MRS Bull. **27**, Feb. 2002.
8. G.R. Barsch, J.A. Krumhansl, L.E. Tanner, and M. Wuttig, Scripta Metall. **21**, 1257 (1987) and references therein.
9. J. A. Krumhansl, in *Lattice Effects in High-T_c Superconductors*, eds. Y. Bar-Yam, T. Egami, J. Mustre de Leon, and A.R. Bishop (World Scientific, Singapore, 1992).
10. S. M. Shapiro, B. X. Yang, Y. Noda, L. E. Tanner, and D. Schryvers, Phys. Rev. B **44**, 9301 (1991).
11. A. N. Lavrov, S. Komiya, and Y. Ando, Nature **418**, 385 (2002); Also see R. Tiwari and V.K. Wadhawan, Phase Trans. **35**, 47 (1991).
12. S. Sergeenkov and M. Ausloos, Phys. Rev. B **52**, 3614 (1995).
13. T. Egami, J. Low Temp. Phys. **105**, 791 (1996).
14. L. Yiping, A. Murthy, G. C. Hadjipanayis, and H. Wan, Phys. Rev. B **54**, 3033 (1996).
15. O. Tikhomirov, H. Jiang, and J. Levy, Phys. Rev. Lett. **89**, 147601 (2002).
16. U. Stuhr, P. Vorderwisch, V. V. Kokorin, and P. A. Lindgard, Phys. Rev. B **56**, 14360 (1997).
17. Y. Murakami, D. Shindo, K. Oikawa, R. Kainuma, and K. Ishida, Acta Mater. **50**, 2173 (2002).
18. Y. Kishi, M. de Graef, C. Craciunescu, T.A. Lograsso, D.A. Neumann, and M. Wuttig, Proceedings of ICOMAT'02, Helsinki, Finland, in press (2002).
19. R. E. Newnham, V. Sundar, R. Yimnirun, J. Su, and Q. M. Zhang, Ceramic Trans. **88**, 15 (1998).
20. J. Yang and D. T. Cassidy, J. Appl. Phys. **77**, 3382 (1995).
21. D. Andrault, G. Fiquet, M. Kunz, F. Visocekas, and D. Hausermann, Science 278, 831 (1997).
22. A. E. Rubin, Meteoritics & Planetary Sci. **32**, 231 (1997).
23. G. B. Olson and H. Hartman, J. Physique Colloque (Paris) C4 **43**, 12, C4-855 1982); S. Celotto and R. C. Pond, Proc. of Intern. Conf. on Martensitic Transformations, ICOMAT-02 (Helsinki, Finland, 2002).
24. S. S. Hecker, MRS Bull. **26**, 672 (2001).
25. C. Manolikas and S. Amelinckx, Phys. Stat. Sol. (a) **60**, 607 (1980); *ibid.* **61**, 179 (1980).
26. A. Planes and L. Mañosa, Solid State Phys. **55**, 159 (2001).

27. L. E. Cross, Ferroelectrics **76**, 241 (1987).
28. C.L.M.H. Navier, *Résumé des Leçons sur l'Application de la Mécanique*, 3ème edition avec des notes et des appendices par A.J.C. Barré de Saint-Venant (Dunod, Paris, 1864).
29. D.S. Chandrasekharaiah and L. Debnath, *Continuum Mechanics* (Academic, San Diego, 1996) p. 218; S.Timoshenko, *History of Strength of Materials* (McGraw-Hill, New York, 1953) p. 229; E.A.H. Love, *A Treatise on the Mathematical Theory of Elasticity* (Dover, New York, 1944); L.E. Malvern, *Introduction to the Mechanics of a Continuous Medium* (Prentice Hall, New Jersey, 1969); I. S. Sokolnikoff, Mathematical Theory of Elasticity (McGraw-Hill, New York, 1946).
30. M. Baus and R. Lovett, Phys. Rev. Lett. **65**, 1781 (1990); Phys. Rev. Lett. **67**, 406 (1991); Phys. Rev. A **44**, 1211 (1991).
31. D. M. Hatch, T. Lookman, A. Saxena, and S. R. Shenoy, Phys. Rev. B, submitted (2002).
32. K. Aizu, J. Phys. Soc. Japan **27**, 387 (1969).
33. S. R. Shenoy, T. Lookman, A. Saxena, and A.R. Bishop, Phys. Rev. B **60**, R12537 (1999).
34. T. Lookman, S. R. Shenoy, K. Ø. Rasmussen, A. Saxena, and A. R. Bishop, Phys. Rev. B, in press.
35. D. M. Hatch and H. T. Stokes, Phys. Rev. B **30**, 5156 (1984).
36. H. T. Stokes and D. M. Hatch, Isotropy Subgroups of the 230 Crystallographic Space Groups (World Scientific, Singapore, 1988). The software package ISOTROPY is available at http://www.physics.byu.edu/~stokesh/isotropy.html, *ISOTROPY* (1991).
37. S. Kartha, J.A. Krumhansl, J.P. Sethna, and L.K. Wickham, Phys. Rev. B **52**, 803 (1995).
38. K. Ø. Rasmussen, T. Lookman, A. Saxena, A.R. Bishop, R.C. Albers, and S.R. Shenoy, Phys. Rev. Lett. **87**, 055704 (2001). The six compatibility equations are actually in two sets of 3 equations (see L.E. Malvern in Ref. 29).
39. See, e.g., J. D. Jorgensen, B. W. Veal, W. K. Kwok, G. W. Crabtree, A. Umezawa, L. J. Nowicki, and A. P. Paulikas, Phys. Rev. B **36**, 5731 (1987).
40. A. Asamitsu, Y. Moritomo, R. Kumai, Y. Tomioka, and Y. Tokura, Phys. Rev. B **54**, 1716 (1996).
41. A. R. Bishop, T. Lookman, A. Saxena, and S. R. Shenoy, preprint (2002). S. R. Shenoy et al., in this volume.
42. K. H. Ahn and A. J. Millis, Phys. Rev. B **64**, 115103 (2001).
43. J. C. Phillips and J. Jung, Phil. Mag. B **81**, 745 (2001).
44. S. J. L. Billinge and P. M. Duxbury, Phys. Rev. B **66**, 064529 (2002).

Lattice and electronic instabilities in oxides

Annette Bussmann-Holder

Max-Planck-Institute for Solid State Research, Heisenbergstr. 1, D-70569 Stuttgart, Germany

ABSTRACT

The rich variety of physical properties observed in perovskite related oxides has been the subject of intensive research for many decades. The discovery of high temperature superconductivity in oxide cuprates has increased the interest in this field even more. Microscopic models for these complex systems mostly deal with strong electronic correlations and rarely address effects arising from the interplay between electronic and lattice degrees of freedom. It will be shown here that these latter effects are of crucial importance in understanding the physical properties of these systems. Especially unconventional electron-phonon interactions are introduced which arise from nonlinear polarizability effects of the oxygen ion O^{2-}.

Perovskite type oxides show an enormous richness in ground state properties, which can be varied by doping, pressure and temperature. Metal-insulator transitions with large isotope effects are observed as well as ferro- and antiferromagnetic properties, ferroelectricity, antiferroelectricity and superconductivity. Charge density wave instabilities and spin density wave formation are also frequently found in these complex compounds, where unusual isotope effects have frequently been reported. Even though quite generally the magnetic properties suggest that electronic correlations play an important role, especially in superconducting oxide cuprates, electron-phonon interactions cannot be ruled out to be of equivalent importance, since strong lattice anomalies are correlated with the onset of superconductivity. On the other hand ferro- and antiferroelectric perovskites are nominally ionic systems where electron-lattice interactions are supposed to be irrelevant. Yet also in these compounds subtle electronic redistributions [1] take place at the structural phase transition which arise from unusual electron-phonon interactions.

An understanding of the microscopic mechanism behind these different instabilities has been suggested by phenomenological considerations which address the configurational instability of the oxygen ion $2p^6$ state [2]. As a free ion state this configuration is unstable and in a crystal a partial stabilization is achieved through the Madelung potential [3]. That this potential leads to a marginal stabilization only is reflected in the fact that the oxygen ion polarizability is strongly temperature, volume and pressure dependent [2]. The origin of this anomaly can be related to the electronic wave functions of the $2p^6$ state which delocalise with increasing oxygen ion radius (Fig. 1) opposite to the isoelectronic configuration of the F^- ion which undergoes a rigid shift with increasing volume to converge at the free ion value. If one subtracts the the charge densities of the two radii a charge transfer into interstitial regions is evidenced which corresponds to hybridisation and dynamical covalency. This implies that also phonon displacements crucially affect the oxygen ion polarizability and that electron – lattice interactions gain giant importance. These interactions cannot be of conventional harmonic form, since the temperature and volume dependence of the polarizability require extensions which are best taken into account by considering multiphonon – density – density interaction terms. A reasonable form for the oxygen ion energy is thus given by: $E(O^{2-}) = \frac{\alpha}{r^6} - \frac{xe^2}{r^2} + \frac{\beta}{r^4}$ where the first term accounts for the volume dependence, the second for the Madelung potential and the last one for the temperature dependence. r is here the ionic radius and the coefficients α, β are determined by the requirement that the energy is a minimum at the frequently observed radius r_0 of 1.42 A. This defines two minima in E as function of r : $r_{1,2}^2 = r_0^2[1 \pm \{1 - \frac{3\alpha}{\beta^2}\}^{1/2}]$ and implies that the oxygen ion energy is degenerate with respect to a double-well position. Since the energy of the oxygen ion $2p^5$ state is lower than the one of the $2p^6$ state, hole doping has completely different consequences for oxides than electron doping.

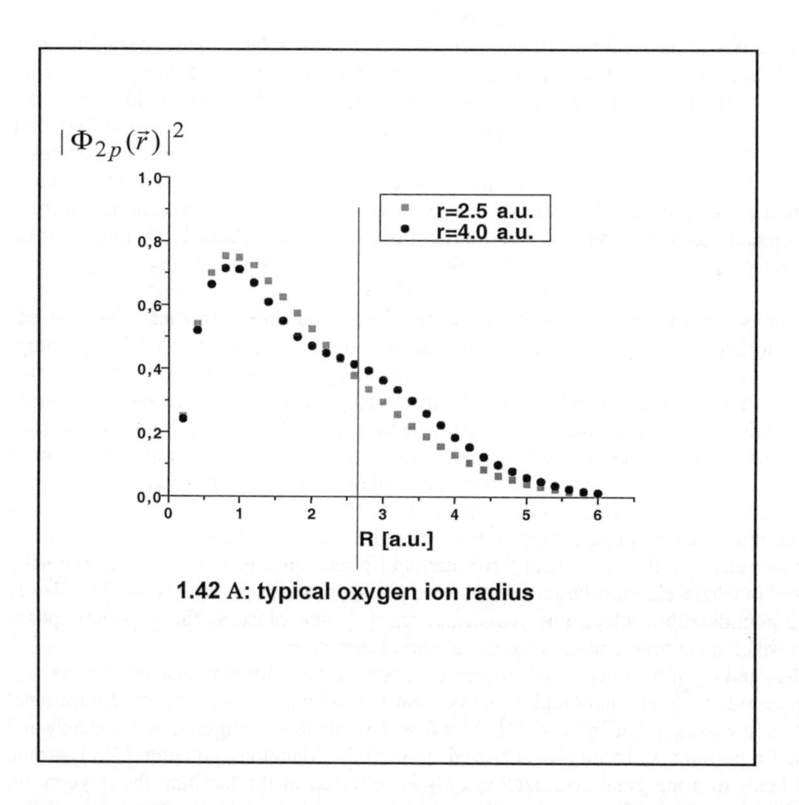

Figure 1: Electronic charge distribution of the oxygen ion O^{2-} for two different radii. The line indicates the typical oxygen ion radius of 1.42 A.

A phenomenological description of this situation has first been given by Migoni et al. [4] who modelled perovskite oxides by a nonlinear shell model representation. The crucial innovation in this model is a nonlinear electron-ion coupling at the oxygen ion lattice site which is anisotropic and with respect to A and B sublattices in the ABO_3 structure. While the coupling to the A ion is harmonic the one to the transition metal ion B consists of an attractive harmonic and a repulsive fourth order term. This model was the first to reproduce, in quantitative agreement with experimental data, temperature dependent properties of ferroelectric perovskites by using a self-consistent phonon approximation, which corresponds to a cumulant expansion of the nonlinear term [5-11]. As has been shown in Refs. 12, 13 the phenomenological model is equivalent a density-density multiphonon interaction model, which – as a consequence for superconductivity – yields a mulitband anharmonicity dominated electron-phonon interaction mechanism [14, 15]. It is important to note here that the nonlinear shell model differs substantially from typical Φ_4 models due to nonlinear interaction terms in the electron-ion coupling [16]. Completely new solutions have been discovered here which introduce new length and time scales, superstructures and local structural anomalies. In addition new mass dependences of lattice modes and the overall lattice potential have been found which explain many of the unusual isotope effects observed in perovskite oxides. Exact nonlinear solutions of the model reveal that the oxygen ion displacement pattern exhibits extra dynamics on top of the phonon dynamics which can be related to stripe formation as e.g. observed in the cuprates [17-19]. Also molecular dynamics studies showed that preformed polarized clusters appear in ferroelectrics which obey different time scales than soft modes and induce a competition between displacive and order/disorder dynamics [20}. Many of the above mentioned findings can be related to the fact that the

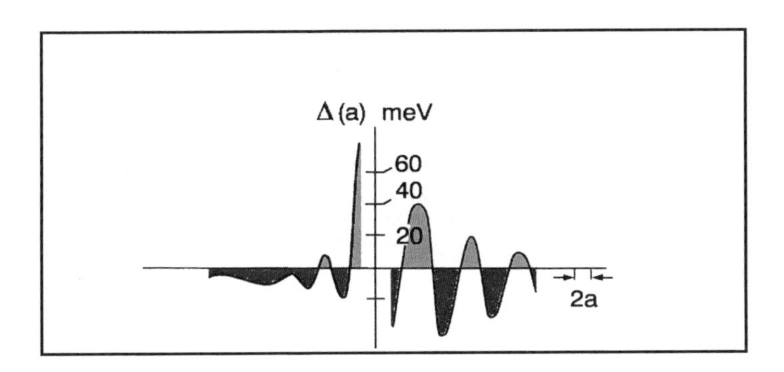

Figure 2: Calculated real space charge ordering on 2-dimensional lattice along the <11> (right side) and the <10> (left side) direction

electron-phonon coupling in this approach is dominated by multiphonon density - density interactions, which lead to a real space variation in the effective coupling. This – in turn – induces spatial modulations in the single electron band energy which correspond to dynamical incommensurate charge density wave type charge ordering as shown in figure 2.

Since ferroelectric perovskites are ionic compounds and undoped superconducting cuprates half band filled Mott-Hubbard insulators, it is at a first glance not obvious that they related to each other. Yet there are a variety of common features as e.g. the perovskite type building units, high dielectric constants [21], the pseudo insulating properties and unusual local structural anomalies [22]. In addition dielectric transitions have been reported in the superconducting compounds [21] which are analogous to the dielectric anomalies observed in ferroelectrics. Also unexpectedly large polarizabilities have been observed which can be related to the oxygen ion. There are – however – also a variety of properties of cuprates which are in strong contrast to ferroelectrics. The cuprates are – when doped – metals, even if their metallic state seems to be unconventional. Many of their physical properties are characterized by a large anisotropy between in-plane and out of plane elements, and their antiferromagnetic insulating behaviour at zero or small doping points to a large Hubbard U at the copper site. The superconducting energy gap seems to be d-wave in the copper oxygen planes and antiferromagnetic correlations persist in the superconducting state [23-28]. The superconducting transition temperatures T_c are by far too large to be explained by a conventional phonon mediated BCS type approach, and the quasi-2-dimensional structure related to the CuO_2-planes was frequently taken as evidence that antiferromagnetic correlations are the origin of the high T_c's [29-33]. Consequently many theoretical approaches focused on new and partially quite exotic mechanisms for superconductivity in oxides. That lattice effects have to be incorporated in modelling these compounds became very evident only recently when new angle resolved photoemission data revealed that a kink in the electronic dispersion, common to many of the superconducting cuprates, finds an explanation only in terms of strong electron phonon coupling [34]. Clearly also previous neutron scattering, XANES and X-ray scattering data showed that lattice effects are present [35-39], but convincing evidence to the community could not be achieved. Together with the new ARPES data also the recent measurements of the extremely large isotope effect on the pseudogap formation temperature [40, 41] showed that a pronounced phonon contribution has to be included in the theoretical modelling of cuprates.

The quasi-2-dimensional nature of these materials enables to decompose them in the undoped or underdoped case into two components, i.e. planes and out-of-plane elements

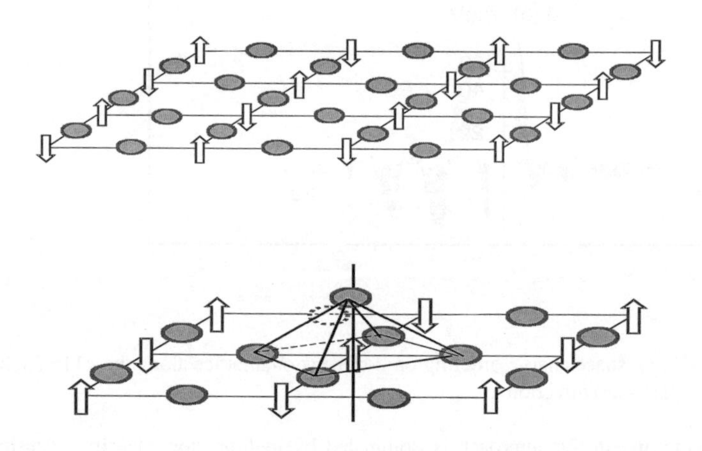

Figure 3: The two components of copper oxides: Upper part: planar component; lower part: out-of-plane component (circles refer to the oxygen ions while arrows indicate the copper position with antiferromagnetic spin alignment).

(Figure 3) Such a two-component scenario has been discussed frequently before [42-46], but the specification of the involved structural elements remained unclear. In the undoped case both components are more or less uncoupled and can be described by the following Hamiltonian [47, 48]:

$$H = H_0 + H_1 + H_2 \qquad (1)$$

with

$$H_0 = \sum_{k_z,\sigma} \xi_{k_z} c^+_{k_z\sigma} c_{k_z\sigma} + \sum_{k_{xy},\sigma} \xi_{k_{xy}} c^+_{k_{xy}\sigma} c_{k_{xy}\sigma}$$

$$H_1 = \sum_{k_z,q} g(q) b_q c^+_{k_z+q} c_{k_z}$$

$$H_2 = \sum_{i,j} T_{xy} n_{xy,i\uparrow} n_{xy,j\downarrow}$$

where in H_0 the kinetic energies ξ of inplane (xy) and out of plane (z) electronic energy bands with creation and annihilation operators c^+, c are given; H_1 is the electron-phonon interaction term which couples to the out-of-plane elements only, with momentum q dependent coupling $g(q)$ and phonon creation and annihilation operators b^+, b. H_2 refers to the strong correlations with interaction T between densities n in the planes. With doping hole localization at the oxygen ion lattice site takes place which lowers the oxygen ion energy. In addition spin singlet formation of the localized hole spin with respect to the nearest neighbour copper ion is the energetically most favourable process, since the corresponding triplet state is some eV higher in energy [49]. In order to compensate for the extra charge dynamical or static plane buckling sets in by means of which new hopping processes are made possible. These in turn induce anharmonic electron-multiphonon processes. The consequence of this scenario is besides strong lattice distortions that the spin next to the hole is ineffective for the antiferromagnetic matrix (figure 4) and a rapid decrease of the antiferromagnetic transition temperature takes place.

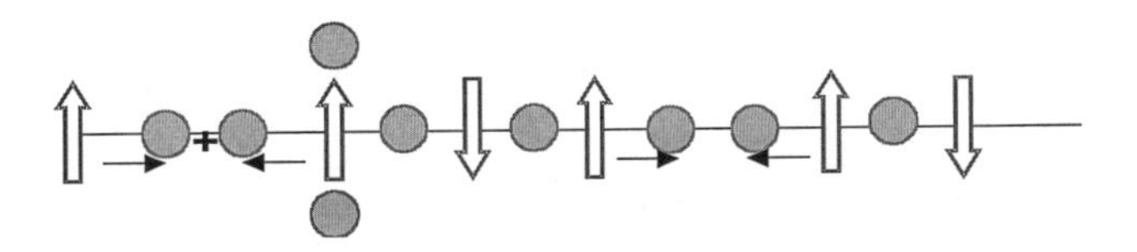

Figure 4: Schematics of the effect of doping. The symbols are the same as in fig. 3, the black arrows indicate the phonon displacements.

The coupling between the two components and the effect of the coupling to the lattice through multiphonon processes changes the Hamiltonian equation 1 to:

$$H = \sum_{i,\sigma} E_{xy,i} c^+_{xy,i,\sigma} c_{xy,i,\sigma} + \sum_{j,\sigma} E_{z,j} c^+_{z,j,\sigma} c_{z,j,\sigma} + \sum_{i,j,\sigma,\sigma'} T_{xy,z} [c^+_{xy,i,\sigma} c_{z,j,\sigma'} + h.c.]$$
$$+ \sum_{i,j} \tilde{T}_{xy} n_{xy,i\uparrow} n_{xy,j\downarrow} + \sum_{i,j,\sigma,\sigma'} \tilde{V}_C n_{i,\sigma} n_{j,\sigma'} + \sum_{i,j,\sigma,\sigma'} V_{pd} n_{i,\sigma} n_{j,\sigma'}$$

(2)

where all energy scales are renormalized through the interaction with the lattice [For details of the renormalization see Ref. 47]. The most important renormalization enters in the single particle energies ξ, which are replaced by $E=\xi-\Delta^*$, where Δ^* is the phonon induced dynamical pseudogap which varies strongly in real space (Fig. 2). These phonon induced modulations have no analytical analogue, and average (meanfield) values have to be used to investigate their role on superconductivity.

In order to relate the above results to HTSC two bands are assumed to be most effective here (the two-band model has been frequently used in the context of conventional superconductivity, but also in relation to cuprates [45, 50-55]): a highly dispersive one dimensional band along the z-direction and a tight-binding two-dimensional band in the x-y-plane: $\xi_{k_{xy}} = t[\cos(k_x a) + \cos(k_y b)]$ with $a \neq b$ to account for the deviations from cubic symmetry. The interaction between both bands is phonon mediated and possible only through multiphonon processes. The starting assumption is that both bands are not superconducting as long as the interaction between both bands is zero. The Hamiltonian equ. 2 is cast into an effective BCS type form which yields:

$$H = H_0 + H_1 + H_2 + H_{12} \tag{3}$$

$$H_1 = \sum_{k_1 k_1' q} V_1(k_1, k_1') c^+_{k_1+q/2\uparrow} c^+_{-k_1+q/2\downarrow} c_{-k_1'+q/2\downarrow} c_{k_1'+q/2\uparrow}$$

$$H_2 = \sum_{k_2 k_2' q} V_2(k_2, k_2') d^+_{k_2+q/2\uparrow} d^+_{-k_2+q/2\downarrow} d_{-k_2'+q/2\downarrow} d_{k_2'+q/2\uparrow}$$

$$H_{12} = \sum_{k_1 k_2 q} V_{12}(k_1, k_2) \{ c^+_{k_1+q/2\uparrow} c^+_{-k_1+q/2\downarrow} d_{-k_2+q/2\downarrow} d_{k_2+q/2\uparrow} + h.c. \}$$

Where the pairing potentials are $V_i(k_i, k_i')$ are given in factorised form like $V_i(k_i, k_i') = V_i \varphi_{k_i} \psi_{k_i}$ with $\varphi_{k_i} = \cos(k_x a) + \cos(k_y b) = \gamma_{k_i}$ for extended s-wave paring and $\varphi_{k_i} = \cos(k_x a) - \cos(k_y b) = \eta_{k_i}$ for d-wave pairing. With the definition $\Delta_{k_i} = \sum_{k_i} V_i(k_i, k_i') < c^+_{k_i\uparrow} c^+_{k_i\downarrow} >$ the reduced Hamiltonian in the absence of the pseudogap reads:

$$H_{red} = \sum_{k_1 \sigma} \xi_{k_1} c^+_{k_1\sigma} c_{k_1\sigma} + \sum_{k_2 \sigma} \xi_{k_2} d^+_{k_2\sigma} d_{k_2\sigma} + \overline{H}_1 + \overline{H}_2 + \overline{H}_{12}$$

with

$$\overline{H}_i = \sum_{k_i, k'_i} [\Delta_{k_i} c^+_{k_i\uparrow} c^+_{-k_i\downarrow} + \Delta^*_{k_i'} c_{-k_i'\downarrow} c_{k_i'\uparrow} - V_i(k_i, k_i')] < c^+_{k_i\uparrow} c^+_{-k_i\downarrow} >< c_{-k_i'\downarrow} c_{k_i'\uparrow} > \qquad i = 1,2$$

$$\begin{aligned} \overline{H}_{12} = \sum_{k_1, k_2} [& V_{12}(k_1, k_2) < c^+_{k_1\uparrow} c^+_{-k_1\downarrow} > d_{-k_2\downarrow} d_{k_2\uparrow} + V_{12}(k_1, k_2) < d_{-k_2\downarrow} d_{k_2\uparrow} > c_{k_1\uparrow} c_{-k_1\downarrow} + \\ & V^*_{12}(k_1, k_2) d^+_{k_2\uparrow} d^+_{-k_2\downarrow} < c_{-k_1\downarrow} c_{k_1\uparrow} > + V^*_{12}(k_1, k_2) c_{-k_1\downarrow} c_{k_1\uparrow} < d^+_{k_2\uparrow} d^+_{-k_2\downarrow} > - \\ & V_{12}(k_1, k_2) < c^+_{k_1\uparrow} c^+_{-k_1\downarrow} >< d_{-k_2\downarrow} d_{k_2\uparrow} > - V^*_{12}(k_1, k_2) < c_{-k_1\downarrow} c_{k_1\uparrow} >< d^+_{k_2\uparrow} d^+_{-k_2\downarrow} >] \end{aligned} \tag{4}$$

For constant interactions V the resulting superconducting gaps are momentum independent. More interesting cases are obtained when the above described momentum dependence of V is considered. The results for T_c for momentum independent and d-wave combined with on-site s-wave gap form are shown are compared to each other in figure 5 as function of the interband interaction. As is obvious from figure 5 a strong enhancement of the superconducting transition temperature due to the interband interaction alone is obtained in both cases. An additional nearly doubling of T_c takes place if one intraband interaction is assumed to be of d-wave type. This might be extremely important when cuprates [57, 58] are compared to MgB_2 where in the latter compound a d-wave gap is not observed even though there is accumulating evidence for a two-gap behaviour [58-60]. Obviously the absence of nodes in the gaps limits

T_c here even if the interband interaction is responsible for the large enhancements of it as compared to conventional systems. The temperature dependence of the coupled gaps for given value of V_{12} is shown in figure 6 for the same two cases as in figure 5. The gap to T_c ratios deviate from BCS predictions, but the temperature dependence is close to it. It is important to note that in the case of s+d wave symmetry the s-wave gap experiences a substantial enhancement as compared to the s+s case which means that the interband coupling to the d-wave gap ties the s-wave gap up to higher values.

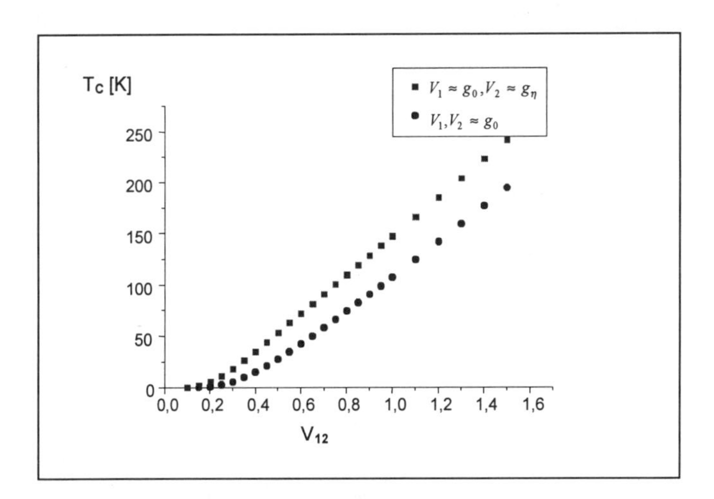

Figure 5: Dependence of T_c on the interband interaction V_{12}. The term g_0 corresponds to a momentum independent intraband interaction.

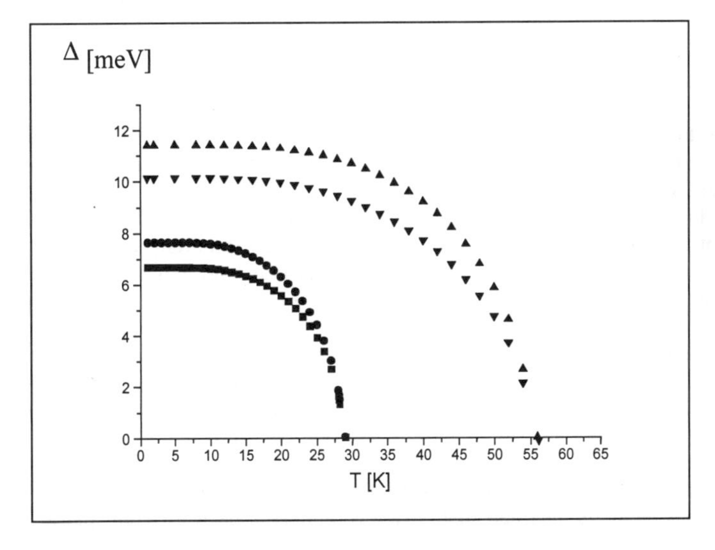

Figure 6: Temperature dependence of the superconducting energy gaps. The two upper curves refer to mixed order paramerters: s+d; the lower ones to onsite isotropic order parameters: s+s.

Until now the important renormalizations due to the coupling to the lattice have been treated implicitly only. The most dramatic effect of the phonons appears, as already mentioned above, in the single particle energies which are renormalized and exhibit a dynamical spatially varying pseudogap Δ*. This corresponds to a level shift in the one-dimensional band along z proportional to $\Delta^* = g(q_0) < b_q^+ > \delta_{q,q_0} 2/\sqrt{N}$ while an additional exponential squeezing has to be included for the two-dimensional band in the x-y-plane [61]: $E_{xy} = [\xi_{k_{xy}} \exp\{| g(q_0)|^2 \coth \frac{\hbar\omega_{q_0}}{2kT}\} - \Delta^*] c_{k_{xy}}^+ c_{k_{xy}}$. Considering the level shift in the one-dimensional band only and concentrating again on the above discussed two cases of pairing potentials, the results as shown in figure 7 are obtained.

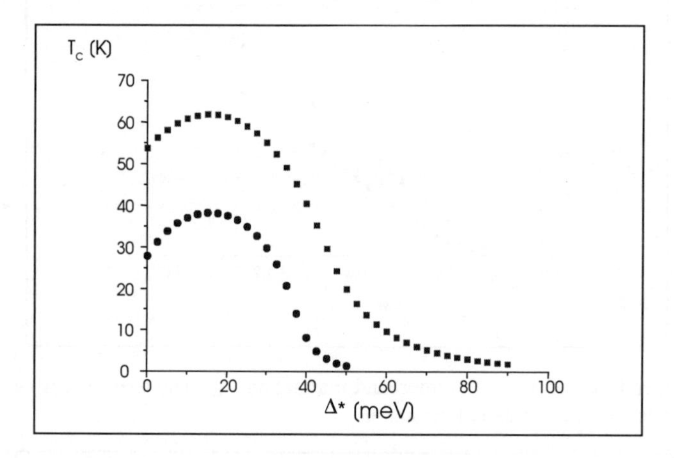

Figure 7: Dependence of T_c on Δ*, where only the band along z is renormalized. Red squares refer to s+d, black circles correspond to s+s pairing symmetry.

For small to intermediate coupling to the lattice, T_c is always enhanced. A sudden drop in T_c takes place if the coupling exceeds a certain strength. Obviously localization sets then in and counteracts superconductivity.

In conclusion, it has been shown that oxygen ion polarizability effects are not only important in ferro- and antiferroelectrics but have also important consequences for high temperature superconductivity, since interband interactions are possible due to multiphonon scattering. In addition the nonlinear anisotropic polarizability induces charge transfer, structural modulations, lattice instabilities and dynamical covalency. The pseudo-two-dimensional structure of the superconducitng cuprates is best modelled by a two-component scenario, where one component refers to z-axis structural elements while the other is given by the CuO_2 planes. The interaction between both subsystems induces superconductivity even if both compounds do not superconduct as long as they uncoupled. Arbitrary large enhancements of T_c can be obtained due to the action of the interband interaction, which most likely leads to a structural instability if it becomes too strong. Another substantial T_c enhancement factor is obtained, if the two components have different pairing potentials, where d-wave pairing for the xy-channel favours superconductivity dramatically. The coupling to the lattice induces a level shift in the single particle energies, which again favours superconductivity up to moderate coupling strengths, but rapidly depresses it due to localization if it becomes too strong.

Acknowledgement: It is a pleasure to acknowledge many fruitful discussions with R. Micnas, K. A. Müller, A. R. Bishop, H. Büttner, A. Simon, T. Egami and A. Bianconi

References

[1] K. H. Weyrich,*Ferroelectrics* **79**, 65 (1988).
[2]G. R. Tessmann, A. H. Kohn, W. Shockley, *Phys. Rev.* **92**, 890 (1953).
[3] A. Bussmann, H. Bilz, R. Roenspiess, K. Schwarz, *Ferroelectrics* **25**, 343 (1980).
[4] R. Migoni, H. Bilz, D. Bäuerle, *Phys. Rev. Lett.* **37**, 1155 (1976).
[5] C. Perry et al., *Phys. Rev. B* **35**, 8666 (1989).
[6] R. L. Migoni, K. H. Rieder, K. Fisher, H. Bilz, *Ferroelectrics* **13**, 377 (1976).
[7] D. Khatib, R. L. Migoni, G. E. Kugel, L. Godefroy, *J. Phys. Condens. Matter* **1**, 9811 (1991).
[8] G. E. Kugel, M. D. Fontana, W. Kress, *Phys. Rev. B* **35**, 813 (1987).
[9] M. Stachiotti, R. Migoni, *J. Phys. Condens. Matter* **2**, 4341 (1990).
[10] M. Stachiotti, R. Magoni, U. Höchli, *J. Phys. Condens. Matter* **3**, 3689 (1991).
[11] A. Bussmann-Holder, H. Bilz, G. Benedek, *Phys. Rev. B* **39**, 9214 (1989).
[12] A. Bussmann-Holder, H. Büttner, A. Simon, *Phys. Rev. B* **39**, 207 (1989).
[13] A. Bussmann-Holder, A. R. Bishop, *Phys. Rev. B* **56**, 5297 (1997).
[14] A. Bussmann-Holder, A. R. Bishop, *Phys. Rev. B* **44**, 2853 (1991).
[15] A. Bussmann-Holder, A. R. Bishop, I. Batistic, *Phys. Rev. B* **43**, 13728 (1991).
[16] G. Benedek, H. Bilz, A. Bussmann-Holder, *Phys. Rev. B* **36**, 630 (1987).
[17] A. Bussmann-Holder, H. Büttner, A. Simon, A. R. Bishop, *Phil. Mag. B* **80**, 1955 (2000).
[18] A. Bussmann-Holder, A. R. Bishop, G. Benedek, *Phys. Rev. B* **53**, 11521 (1996).
[19] A. Bussmann-Holder, H. Büttner, A. R. Bishop, *J. Phys. Cond. Mat.* **12**, L115 (2000).
[20] M. Stachiotti, G. Dobry, R. Magoni, A. Bussmann-Holder, *Phys. Rev. B* **47**, 2473 (1996).
[21] S. Sridhar et al., *Phys. Rev. B* **63**, 92508 (2001).
[22] A. Bianconi, N. L. Saini, A. Zanzara, M. Missori, T. Rossetti, H. Oyanagi, Y. Yamaguchi, K. Oka, T. Ito, *Phys. Rev. Lett.* **76**, 3412 (1996).
[23] H. Mook, Dai Pencheng, A. Hayden, G. Äppli, C. Perring, F. Dogan, *Nature* **395**, 580 (1998).
[24] J. Haase, C. Slichter, R. Stern, F. Milling, D. Hinks, *J. Supercond.: Incorp. Novel Magnetism* **13**, 723 (2000).
[25] J. Zaanen, O. Gunnarson, *Phys. Rev. B* **40**, 7391 (1989).
[26] Y. Yoshizawa, *Phys. Rev. B* **61**, R854 (2000).
[27] Tranquada et al., *Phys. Rev. Lett.* **78**, 338 (1997).
[28] Ichikawa et al., *Phys. Rev. Lett.* **85**, 1738 (2000).
[29] V. J. Emery, S. A. Kivelson, O. Zachar, *Phys. Rev. B* **56**, 6120 (1997).
[30] E. Dagotto, *Rev. Mod. Phys.* **66**, 763 (1994).
[31] D. J. Scalapino, *Phys. Rep.* **250**, 329 (1995).
[32] J. R. Schrieffer, X. G. Wen, S. C. Zhang, *Phys. Rev. B* **39**, 11663 (1989).
[33] D. Pines, *Physica C* **235-240**, 113 (1994).
[34] A. Lanzara et al., *Nature* **412**, 510 (2001).
[35] S. J. L. Billinge, G. H. Kwei, H. Takagi, *Phys. Rev. Lett.* **72**, 2282 (1994).
[36] G. B. Teitel'baum, B. Büchner, H. de Grockel, *Phys. Rev. Lett.* **84**, 2949 (2000).
[37] R. J. McQueeney, Y. Petrov, T. Egami, M. Yethiraj, G. Shirane, Y. Endoh, *Phys. Rev. Lett.* **82,** 628 (1999).
[38] A. Perali, A. Bianconi, A. Zanzara, N. L. Saini, *Sol. St. Comm.* **100**, 181 (1996).

[39] O. Chmaissen, J. D. Jorgensen, S. Short, A. Kizhnik, Y. Eckstein, H. Shaked, *Nature* **397**, 45 (1999).
[40] A. Lanzara, Guo-Meng Zhao, N. L. Saini, A. Bianconi, K. Condor, H. Keller, K. A. Müller, *J. Phys. Cond. Mat.* **11**, L541 (1999).
[41] D. RubioTemprano, J. Mesot, A. Furrer, K. Condor, H. Mutka, K. A. Müller, *Phys. Rev. Lett.* **84**, 1990 (2000).
[42] C. J. Stevens, D. Smith, C. Chen, R. F. Ryan, B. Podobnik, D. Mihailovic, G. A. Wagner, J. E. Evetts, *Phys. Rev. Lett.* **78**, 2212 (1997).
[43] D. Mihailovic, T. Mertelj, K. A. Müller, *Phys. Rev. B* **57**, 6116 (1998).
[44] L. P. Gor'kov, A. V. Sokol, *JETP Lett.* **46**, 420 (1987).
[45] S. Robaszkiewicz, R. Micnas, J. Ranninger, *Phys. Rev. B* **36**, 180 (1987).
[46] D. Mihailovic, K. A. Müller, *High Temperature Superconductivity 1996: 10 Years after the discovery (NATO asi series, E434)* ed. E. Kaldis, E. Liarokapis, K. A. Müller (Kluwer : Dordrecht), p. 243 (1997).
[47] A. Bussmann-Holder, K. A. Müller, R. Micnas, H. Büttner, A. Simon, A. R. Bishop, T. Egami, *J. Phys. Cond. Mat.* **13**, L169 (2001).
[48] A.Bussmann-Holder, A. R. Bishop, H. Büttner, T. Egami, R. Micnas and K. A. Müller *J. Phys.: Cond. Mat.* **13**, L545 (2001).
[49] F. C. Zhang, T. M. Rice, *Phys. Rev. B* **37**, 3759 (1988).
[50] H. Suhl, B. T. Matthias and L. R. Walker, *Phys. Rev. Lett*. **3**, 552 (1959).
[51] V. A. Moskalenko, *Fiz. Metal. Metalloved.* **8**, 503 (1959).
[52] J. Kondo, *Prog. Theor. Phys*. **29**, 1 (1963).
[53] B. T. Geilikman, R. O. Zaitsev and V. Z. Kresin, *Sov. Phys. – Solid State* **9**, 642 (1967).
[54] V. Z. Kresin, *J. Low Temp. Phys.* **11**, 519 (1973).
[55] V. Z. Kresin and S. Wolf, *Physica C* **169**, 476 (1990).
[56] K. A. Müller and H. Keller, in *Proc. NATO ASI „Material aspects of High-T_c superconductivity: 10 years after the discovery*". (Kluwer, 1997).
[57] K. A. Müller, *Phil. Mag. Lett.* **82**, 270 (2002).
[58] Amy Y. Liu, I. I. Mazin and Jens Kortus, *Phys. Rev. Lett.* **87**, 087005 (2001).
[59] H. D. Yang, J.-Y. Lin, H. H. Li, F. H. Hsu, C. J. Liu, S.-C. Li, R. C. Yu and C.-Q. Jin, *Phys. Rev. Lett.* **87**, 167003 (2001).
[60] X.-K. Chen, M. J. Konstantinovi_, J. C. Irwin, D. D. Lawrie and J. P. Franck, *Phys. Rev. Lett.* **87**, 157002 (2001).
[61] I. G. Lang and Yu. A. Firsov, *Soviet Physics JETP* **16**, 1301 (1963).

CONTRASTING PATHWAYS TO MOTT GAP COLLAPSE IN ELECTRON AND HOLE DOPED CUPRATES

R. S. MARKIEWICZ
Northeastern University, 360 Huntington Avenue, Boston MA 02115, USA
E-mail: markiewic@neu.edu

Recent ARPES measurements on the electron-doped cuprate $Nd_{2-x}Ce_xCuO_4$ can be interpreted in a mean field model of uniform doping of an antiferromagnet, with the Mott gap closing near optimal doping. Mode coupling calculations confirm the mean field results, while clarifying the relation between the Mott gap and short-range magnetic order. The *same* calculations find that hole doped cuprates should follow a strikingly different doping dependence, involving instability toward spiral phases or stripes. Nevertheless, the magnetic order (now associated with stripes) again collapses near optimal doping.

1. Introduction

1.1. Mode Coupling Theories

In conventional itinerant ferro- and antiferromagnets, mode-coupling theories[1,2] have proven of value in treating the role of fluctuations in reducing or eliminating long-range order, as well as in the development of local moments, Curie-like susceptibility, and in general the crossover to magnetic insulators. Similar approaches have been applied to charge-density wave systems[3] and the glass transition[4].

Attempts to apply such a formalism to study the antiferromagnetism of the cuprate superconducting compounds have been frustrated, since the antiferromagnetic (AF) phase is found to be unstable against hole doping, toward either an incommensurate AF phase[5] or phase separation[6]. Here, it is demonstrated that this tendency to instability is either absent or greatly reduced in electron-doped cuprates, and the mode coupling analysis can provide a detailed description of the collapse of the Mott gap with doping. The results are of great interest of themselves: the collapse is associated with one or more quantum critical points (QCPs), and superconductivity is optimized close to one QCP. However, the results have an additional importance in the light they shed on the more complicated problem of the *hole doped* cuprates. First, the effective Hubbard U parameter has a significant doping dependence, from $U \sim W$ at half filling (where $W = 8t$ is the bandwidth and t the nearest neighbor hopping parameter) to $U \sim W/2$ near optimal doping. Secondly, the *same* mode coupling theory which works so well for electron-doped cuprates, *breaks down* for hole doping, due to the above noted electronic instability. This suggests that (a) the electron-hole asymmetry must be due to a band structure effect and (b) a suitable generalization of mode coupling theory should be able to incorporate the effects of this instability.

1.2. ARPES of $Nd_{2-x}Ce_xCuO_4$

Recently, Armitage, et al.[7] succeeded in measuring angle-resolved photoemission spectra (ARPES) of $Nd_{2-x}Ce_xCuO_4$ (NCCO) as a function of electron doping, from essentially the undoped insulator $x = 0$ to optimal doping $x = -0.15$. The doping dependence is strikingly different from that found in *hole-doped* $La_{2-x}Sr_xCuO_4$ (LSCO)[8,9]. Both systems start from a Mott insulator at half filling, with ARPES being sensitive only to the lower Hubbard band (LHB), approximately 1eV below the Fermi level E_F. With hole doping, the LHB remains well below E_F while holes are added in mid-gap states (as expected, e.g., in the presence of nanoscale phase separation[10]); for electron doping, the Fermi level shifts to the upper Hubbard band (UHB), and the electrons appear to uniformly dope the antiferromagnet.

Remarkably, the full doping dependence can be simply described by a mean-field (MF) $t-t'-U$ Hubbard model[11], where t' is the second neighbor hopping, U is the onsite coulomb repulsion, and a one band (copper only) model was assumed for simplicity.[a] A key finding is that the Hubbard U is doping dependent, leading to a quantum critical point (QCP) just beyond optimal doping, where the Mott gap vanishes. The appearance of a peak in superconductivity near an AFM QCP is a fairly common occurance[12]; in particular, something similar has been observed[13] in the hole-doped cuprates. However, the MF theory is problematic, in that the Mott gap is associated with long-range Néel order, and the MF model predicts anomalously high values for T_N.

[a] Actually, in Ref. 11 a third neighbor hopping t'' was included to optimize the fit to the experimental Fermi surface curvature. The changes induced by this parameter are small, and it will be ignored in the present calculations.

2. Mode Coupling Theory

This anomalous behavior can be cured by incorporating the role of fluctuations. Indeed, it is known that in a two-dimensional system, the Néel transition can only occur at $T = 0$ – the Mermin-Wagner (MW) theorem[14]. By treating fluctuations within a mode coupling analysis, the MW theorem is satisfied[15], and the Mott gap is completely decoupled from long-range spin-density wave (SDW) order. Even though $T_N = 0$ a large Mott (pseudo)gap is present even well above room temperature near half filling – due to *short-range* AFM order. The MF gap and transition temperature are found to be approximately the pseudogap and T^* the onset temperature for the pseudogap opening.

The calculation can be summarized as follows. In a path integral formulation of the Hubbard model[16], the quartic term is decoupled via a Hubbard-Stratonovich transformation and the fermoinic degrees integrated out. The resulting action is then expanded to quartic order in the Hubbard-Stratonovich fields ϕ. The quadratic interaction reproduces the RPA theory of the Hubbard model – the quadratic coefficient is just $U\delta_{0q}$, where δ_{0q} is the (inverse) Stoner factor $\delta_q = 1 - U\chi_0(\vec{Q} + \vec{q}, \omega)$ (here $\vec{Q} = (\pi, \pi)$ is the wave vector associated with the commensurate SDW). The quartic interaction, parametrized by the coefficient u evaluated at $\vec{Q}$, $\omega = 0$, represents coupling between different magnetic modes. The effects of this term cannot be treated in perturbation theory, and a self-consistent renormalization (SCR) scheme[2] is introduced to calculate the renormalized Stoner factor $\delta_q = \delta_{0q} + \lambda$. A self-consistent equation for δ is found, which can be solved if the band parameters and the interactions U and u are known.

To simplify the calculation, λ is assumed to be independent of $\vec{q}$ and ω, and the Stoner factor is expanded near $\vec{Q}$ as

$$\delta_q(\omega) = \delta + Aq^2 - B\omega^2 - iC\omega, \tag{1}$$

similar to the form assumed in nearly antiferromagnetic Fermi liquid (NAFL)[17] theory. The imaginary term linear in frequency is due to the presence of low energy magnon excitations in the vicinity of the 'hot spots' – the points where the Fermi surface intersects the Brillouin zone diagonal. Here fluctuations toward long-range Néel order lead to strong, Bragg-like scattering which ultimately leads to the magnetic Brillouin zone boundary at T_N. Independent of the parameter values, it is found that $\delta > 0$ for $T > 0$ – the MW theorem is satisfied, while for electron doping $\delta \to 0$ as $T \to 0$ up to a critical doping – there is a QCP associated with $T = 0$ SDW order. For hole doping the calculation breaks down – the parameter A is negative in a significant doping regime, as will be discussed further below.

The key insights of mode coupling theory are:

• The Mott transition is dominated by hot spot physics, which creates zone-edge magnons. The condensation of these magnons creates a new zone boundary, and opens up a gap (the Mott gap) in the electronic spectrum. In two dimensions (2D), there can be no Bose condensation at finite temperatures, but the pileup of lowest energy magnons as T decreases leads to the appearence of a Mott pseudogap and a T=0 transition to long-range SDW order.

• Evidence for the existence of local magnons comes from well-defined *plateaus* in the spin susceptibility. Plateaus are seen in (a) the doping dependence of the susceptibility at $\vec{Q}$, (b) the $\vec{q}$ dependence of the susceptibility near $\vec{Q}$, and (c) the ω dependence of the susceptibility (both real and imaginary parts) at $\vec{Q}$. These plateaus introduce *cutoffs* in the $\vec{q}$ and ω dependence of the Stoner factor, Eq. 1, which in general *cannot* be sent to ∞, in contrast to the NAFL model. Also, the flatness of the plateau tops makes it difficult to estimate the model parameters from first principles. In particular A is strongly temperature dependent, while u is anomalously small (this problem had been noted previously[18]).

• A finite Néel temperature can be generated by interlayer coupling. In the cuprates, such coupling is typically frustrated, and Néel order more likely arises from a Kosterlitz-Thouless transition, after the spin dimensionality is reduced by, e.g., spin-orbit coupling[19].

3. Results

3.1. Susceptibility

In analyzing the ARPES data, a tight-binding band is assumed,

$$\epsilon_k = -2t(c_x + c_y) - 4t' c_x c_y, \tag{2}$$

with $c_i = \cos k_i a$, $t = 0.326eV$, and $t'/t = -0.276$. The Hubbard U is doping dependent, $U/t = 6$, 5, 3.5, and 2.9 at $x = 0$, -0.04, -0.10, and -0.15, respectively, and the mode coupling constant is adjusted to reproduce the low-temperature spin stiffness at half filling[20], $u^{-1} = 0.256eV$. The bare susceptibility

$$\chi_0(\vec{q}, \omega) = -\sum_{\vec{k}} \frac{f(\epsilon_{\vec{k}}) - f(\epsilon_{\vec{k}+\vec{q}})}{\epsilon_{\vec{k}} - \epsilon_{\vec{k}+\vec{q}} + \omega + i\delta}, \tag{3}$$

is evaluated near $\vec{Q}$ to determine the parameters of Eq. 1.

The susceptibility $\chi_0(\vec{Q}, 0)$ has approximately the shape of a plateau as a function of doping, Figure 1a, bounded by the critical points x_H and x_C

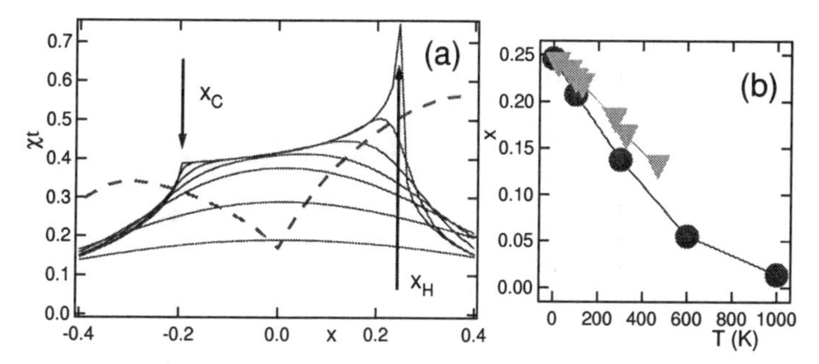

Figure 1. **(a) Susceptibility χ_0 at $\vec{Q}$ as a function of doping for several** temperatures. From highest to lowest curves near $x = 0.1$, the temperatures are $T = 1$, 100, 300, 600, 1000, 2000, and 4000 K. Dashed line $= 1/U_{eff}$. (b) Circles = pseudo-VHS (peak of χ_0) as a function of temperature T_V; triangles $= T_{incomm}$.

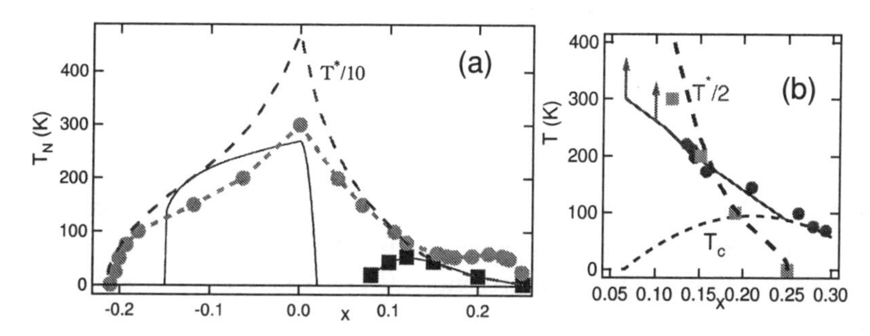

Figure 2. (a) Mean-field transition temperature T^* (long dashed line) compared with Néel temperature of NCCO and LSCO (solid line) plus magnetic transition temperature of stripe ordered phase in Nd-substituted LSCO[25], Circles = model calculation of T_N assuming unfrustrated interlayer hopping. (b) Comparison of mean-field transition (long dashed line) to various estimates of pseudogap temperature: from photoemission[22] (solid line), heat capacity[23] (squares) and tunneling[24] (circles $= \Delta/3$, with Δ the tunneling gap). Short dashed line = superconducting T_c.

where the Fermi surface ceases to have hot spots. These special points act as *natural phase boundaries* for antiferromagnetism, due to the sharp falloff in χ off of the plateau.

The special point x_H coincides at $T = 0$ with the Van Hove singularity (VHS) of the band. Remarkably, the susceptibility peak has a strong temperature dependence[21], defining a pseudo-VHS; Fig. 1b shows the temperature $T_V(x)$, at which the susceptibility peaks at x. This can be understood by noting that the energy denominator of χ, Eq. 3, is *independent of* t', and thus would lead to a large peak at half filling, x=0 (associated with states along the zone diagonal). At low temperatures, the difference in Fermi functions in the numerator cuts this off, and forces the peak to coincide

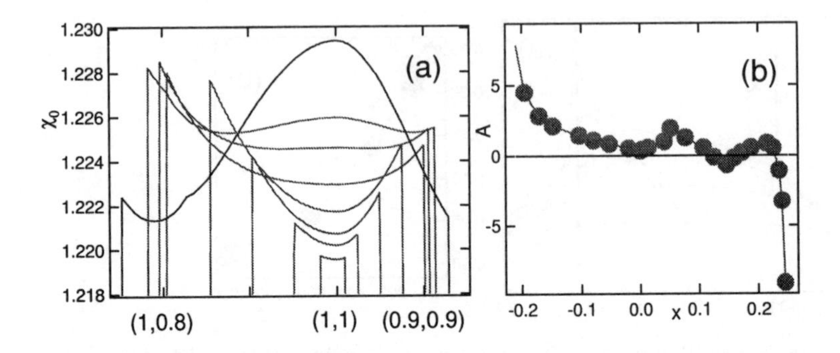

Figure 3. (a) Expanded view of susceptibility χ_0 on the plateaus near $\vec{Q}$ for a variety of dopings at $T = 1K$. From highest to lowest curves near $\vec{Q}$, the chemical potentials are μ = -0.25, -0.20, -0.15, -0.10, -0.05, and -0.02eV. All curves except $\mu = -0.20eV$ have been shifted vertically to fit within the expanded frame. (b) Calculated $A(x)$.

with the VHS. As T increases, more states near the zone diagonals become available, causing the peak susceptibility to shift towards half filling. For a temperature T_{incomm} slightly above T_V, the curvature A becomes negative, signalling the instability of the commensurate AFM state.

The dashed line in Fig. 1a represents $1/U_{eff}$, where U_{eff} is a doping dependent Hubbard U, estimated from a screening calculation[11]. The intersection of the dashed line with one of the solid lines defines the mean-field Néel transition, $\chi_0 U_{eff} = 1$, Fig. 2a (long-dashed line). As will be shown below, once fluctuations are included, the mean field transition turns into a pseudogap onset T^*, while the actual onset of long range magnetic order is suppressed to much lower temperatures. From Fig. 2b, it can be seen that the mean-field T^* is consistent with a number of estimates[22,23,24] of the experimental pseudogap for *hole* doping, while a simple calculation of the three-dimensional Néel transition associated with interlayer coupling (circles in Fig. 2a – see the Appendix for details) can approximately reproduce the experimental observations (solid lines) – if the transitions associated with magnetic order on quasistatic stripes[25] are included (squares).

In addition to plateaus in *doping*, the hot spots lead to plateaus in the frequency and wave number dependence of $\chi_0(\vec{q},\omega)$. For instance, Fig. 3a shows plateaus in $\chi_0(\vec{Q}+\vec{q},0)$ at a series of dopings at $T = 1K$. Once χ_0 is known the parameters A and C of Eq. 1 can be calculated; B is quite small, and can in general be neglected. The plateau width $q_c \rightarrow 0$ as $x \rightarrow x_C$, leading to a strong T-dependence of A, Fig. 3b.

Given the model parameters, the self-consistent equation for δ becomes[15]

$$\delta = \delta_0 + \frac{3ua^2}{\pi^2 C}\int_0^{q_c^2} dq'^2 \int_0^{\alpha_\omega} dx coth(\frac{x}{2CT})\frac{x}{(\delta + Aq'^2)^2 + x^2}. \quad (4)$$

which has the approximate solution

$$Z\delta - \bar{\delta}_0 = \frac{3ua^2 T}{\pi A} ln(\frac{2CT}{e\delta}), \quad (5)$$

where $\bar{\delta}_0 = \delta_0 + \eta - 1$,

$$\eta = 1 + \frac{3uq_c^2 a^2}{\pi^2 C}(\frac{1}{2}ln[1 + a_q^{-2}] + \frac{tan^{-1}(a_q)}{a_q}), \quad (6)$$

$a_q = Aq_c^2/\alpha_\omega$, and q_c and α_ω are wavenumber and (normalized) frequency cutoffs, respectively, and the exact form of Z is not required. From the logarithm in Eq. 5, δ must be greater than zero for all $T > 0$, so there is no finite temperature phase transition. At low temperatures, the δ on the left hand side of Eq. 5 can be neglected, leading to a correlation length

$$\xi^2 = \frac{A}{\delta} = \xi_0^2 e^{4\pi\rho_s/T}, \quad (7)$$

with $\xi_0^2 = eA/2TC$ and

$$\rho_s = \frac{\pi A|\bar{\delta}_0|}{24ua^2}. \quad (8)$$

Here, $\bar{\delta}_0$ is the quantum corrected Stoner factor, *which controls the* $T = 0$ *QCP*: there is long-range Néel order at $T = 0$ whenever $\bar{\delta}_0 \leq 0$, or $U\chi_0 \geq \eta$.

3.2. ARPES Data

From the susceptibility, the contribution to the electronic self energy due to one magnon scattering can be calculated. The imaginary part of the susceptibility can be written

$$Im\Sigma(\vec{k},\omega) = \frac{-g^2\chi_0}{V}\sum_{\vec{q}}\int_{-\alpha_\omega/C}^{\alpha_\omega/C} d\epsilon[n(\epsilon) + f(\xi_{\vec{k}+\vec{q}})] \times$$

$$\times\delta(\omega + \epsilon - \xi_{\vec{k}+\vec{q}})\frac{C\epsilon}{(\delta + Aq'^2)^2 + (C\epsilon)^2}. \quad (9)$$

where the coupling is approximately $g^2\chi_0 \simeq 3U/2$. Since ϵ is peaked near zero when $\vec{q} \simeq \vec{Q}$, $Im\Sigma$ is approximately a δ-function at $\xi_{\vec{k}+\vec{Q}}$. Approximating $Im\Sigma = -\pi\bar{\Delta}^2\delta(\omega - \xi_{\vec{k}+\vec{Q}})$, then

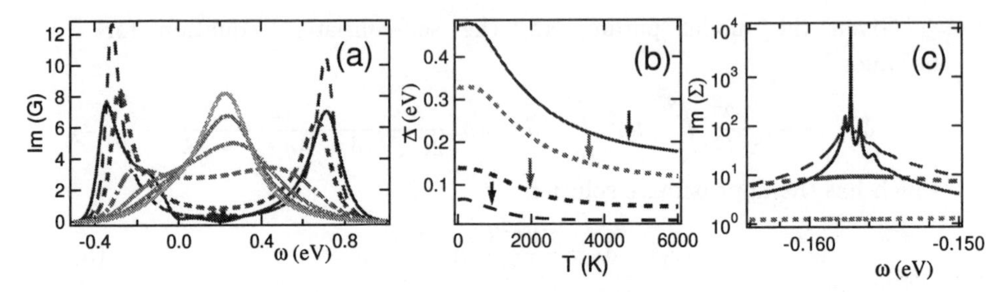

Figure 4. (a)Spectral function for $x = 0.0$, $\vec{k} = (\pi/2, \pi/2)$, at $T = 100$, 500, 1000, 2000, 3000, 4000, and 5000K (larger splittings corresponding to lower T's). (b) $\bar{\Delta}(T)$; solid line: $x = 0.0$, dotted lines: $x = -0.04$, short-dashed lines: $x = -0.10$, long-dashed lines: $x = -0.15$. (c)$Im(\Sigma)$ for $x = 0$, $\vec{k} = (\pi, 0)$ at $T = 100$ (solid line), 500 (long dashed line), 1000 (short dashed line), and 2000K (dotted line).

The importance of this result can be seen by noting that, by Kramers-Kronig,

$$Re\Sigma(\vec{k}, \omega) = \frac{\bar{\Delta}^2}{\omega - \xi_{\vec{k}+\vec{Q}}}, \tag{11}$$

so that

$$G^{-1}(\vec{k}, \omega) = \omega - \xi_{\vec{k}} - Re\Sigma^R(\vec{k}, \omega) = \frac{(\omega - \xi_{\vec{k}})(\omega - \xi_{\vec{k}+\vec{Q}}) - \bar{\Delta}^2}{\omega - \xi_{\vec{k}+\vec{Q}}}. \tag{12}$$

The zeroes of G^{-1} are identical to the mean field results for long-range AFM order[26,11], except that the long-range gap $\Delta = U < m_z >$ is replaced by the *short-range* gap $\bar{\Delta} \sim U\sqrt{< m_z^2 >}$.

Figure 4a shows the spectral function $A(\vec{k}, \omega) = Im(G(\vec{k}, \omega))/\pi$ for $x = 0$ at $\vec{k} = (\pi/2, \pi/2)$ at a series of temperatures. The spectrum is split into upper and lower Hubbard bands, with a gap approximately $2\bar{\Delta}$. The short-range order gap $\bar{\Delta}$ is plotted in Fig. 4b; it is finite for all temperatures, but increases significantly close to the mean-field Néel temperature (arrows). The net dispersion for two dopings, $x = 0$, -0.15, is shown in Fig. 5; it is in good agreement with the experimental[7] and mean-field[11] results. The build up of hot spot magnons is reflected in the growth of $Im(\Sigma(\vec{k}, \omega))$ near $\omega = \xi_{\vec{k}+\vec{Q}}$, Fig. 4c (note the logarithmic scale).

4. Implications for Hole Doping

One expects, and observes, a certain degree of symmetry between electron and hole doping: there is a susceptibility plateau associated with hot spots, Fig. 1, which terminates near optimal doping; the termination of magnetic order in electron-doped cuprates at a QCP near optimal doping is matched

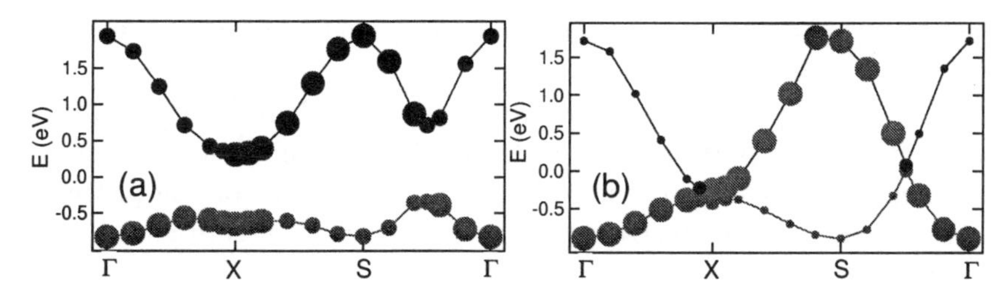

Figure 5. Dispersion relations for (a) x=0 and (b) x=-0.15, at T=100K. Brillouin zone directions are $\Gamma = (0,0)$, $X = (\pi, 0)$, $S = (\pi, \pi)$.

in hole-doped cuprates by the observation of a QCP near optimal doping which appears to be associated with loss of magnetic correlations[13]; and in both cases an optimal, probably d-wave superconductivity is found near the QCP. Also, it is expected that U would have a similar decrease with doping for either electron or hole doping (see the model calculation, dashed line in Fig. 1).

On the other hand, there are also significant differences, not the least of which is the *magnitude* of the superconducting T_c. Most significantly, there is considerable evidence for the appearence of nanoscale phase separation – either in the form of stripes[27] or blobs[28] – for hole doped cuprates, while the evidence is weaker or absent for electron doping[29,30]. As noted above, this difference arises naturally within the present calculations, which find instability of the commensurate magnetic order for hole doping only (e.g., Fig. 3).

Given these similarities and differences, the present calculations can shed some light on aspects of hole-doped physics:

(1) A large *pseudogap* is present at half filling, associated with short-range magnetic order. Just as for electron doping, it should persist with hole doping until short-range magnetic order is lost. The observed[13] connection of the loss of magnetic fluctuations with the collapse of the pseudogap near $x = 0.19$ strongly suggests an identification of the observed pseudogap at T^* with the Mott pseudogap, as found for electron doping. While a number of theories have proposed that the pseudogap is associated with superconducting fluctuations, these seem to turn on at a temperature lower than T^*.[31]. Remarkably, despite the complications associated with stripes, the mean-field transition temperature is within a factor of two of the observed pseudogap temperatures, Fig. 2b.

(2) The present calculations point to a close connection between the Van Hove singularity (VHS) and the instability of commensurate magnetic order, Fig. 1b. This strongly suggests that the VHS is responsible for the

asymmetry between electron and hole doping, and in particular for any frustrated phase separation.

(3) If the stripes are associated with frustrated phase separation, it is important to identify the second (metallic) phase and understand what interaction stabilizes it (particularly since this is likely to be the phase in which high-T_c superconductivity arises). In a purely Hubbard model, this would be a ferromagnetic phase, hence probably incompatible with superconductivity. However, the reduction of U with doping found here strengthens the case for a nonmagnetic charge stripe associated with interactions beyond the Hubbard model[32].

5. Discussion: Polaron Limit

In the very low doping limit, isolated charge carriers should form (spin or charge) polaronic states for either sign of doping. The asymmetry between hole and electron doping would then be reflected in interpolaronic interactions being strongly attractive for hole doping. For electron doping, the isolated polarons could be much more easily pinned in the AFM background, leading to the much stronger localization found in NCCO[33].

Acknowledgments

This work has been supported in part by the Spanish Secretaria de Estado de Educación y Universidades under contract n^o SAB2000-0034. The work was carried out while I was on sabbatical at the Instituto de Ciencia de Materiales de Madrid (ICMM) in Madrid, and the Laboratory for Advanced Materials at Stanford. I thank Paco Guinea, Maria Vozmediano, and Z.X. Shen for inviting me, and for numerous discussions.

Appendix: c-Axis Coupling

A toy model is introduced to study the effect of interlayer coupling on generating a finite Néel transition temperature T_N. The interlayer hopping is assumed to be a constant t_z independent of in-plane momentum. While a term of the form $t_z(c_x - c_y)^2$ would not greatly change the results, in the physical cuprates alternate CuO_2 planes tend to be *staggered*, which should lead to frustration $t_z(\vec{Q}) = 0$, and greatly reduced interlayer coupling. Indeed, in the cuprates it is entirely possible that interlayer coupling is negligible, and that the Néel transition is actually of Kosterlitz-Thouless form, due to reduced spin dimensionality caused by spin-orbit coupling[19].

Nevertheless, it is instructive to see how constant-t_z interlayer coupling can generate a finite T_N. The revised Eq. 5 can be written in the symbolic form

$$Z\delta = \bar{\delta}_0 + \frac{T}{T_0} ln(\frac{D_0}{D + 2\delta}), \tag{A.13}$$

where $T_0 = \pi^2 A/6ua^2$ and $D_0 = 4CT/e$ are (doping-dependent) constants (Eq. 5 – the extra $\pi/2$ in T_0 coming from the q_z-integral) and[34] $D \propto t_z^2$. Thus for finite t_z, there is a non-zero T_N given by the solution of $\bar{\delta}_0 + T/T_0^* = 0$, with $T_0^* = T_0/\ln(D_0/D)$. For the calculation in Fig. 2a, a constant $T_0^* = 1200K$ was assumed, but it is interesting to note that when the correct doping dependence of the parameters is included, $T_N \to 0$ as $A \to 0$, suggesting that the much steeper falloff of T_N with hole doping is related to phase separation.

Equation A.13 can be rewritten using Eq. 8. The Néel transition occurs when

$$J_z[\xi(T_N)/\xi_0(T_N)]^2 = \Gamma T_N, \tag{A.14}$$

with $J_z/J = (t_z/t)^2$, $\Gamma = 16C/edU$, $d = D/t_z^2$ and $J = 4t^2/U$, suggestive of a form of interlayer coupling proposed earlier[35].

References

1. K. K. Murata and S. Doniach, Phys. Rev. Lett. **29**, 285 (1972).
2. T. Moriya, "Spin Fluctuations in Electron Magnetism", (Springer, Berlin, 1985).
3. P.A. Lee, T.M. Rice, and P.W. Anderson, Sol. St. Commun. **14**, 703 (1974); N. Suzuki and K. Motizuki, in "Structural Phase Transitions in Layered Transition Metal Compounds", ed. by K. Motizuki (Reidel, Dordrecht, 1986), p. 135; R.S. Markiewicz, Physica C**169**, 63 (1990); S. Andergassen, S. Caprara, C. Di Castro, and M. Grilli, Phys. Rev. Lett. **87**, 056401 (2001).
4. W. Götze, J. Phys.: Cond. Matt **11**, A1 (1999).
5. B.I. Shraiman and E.D. Siggia, Phys. Rev. Lett. **62**, 1564 (1989).
6. C. Zhou and H.J. Schulz, Phys. Rev. B**52**, 11557 (1995).
7. N.P. Armitage, D.H. Lu, C. Kim, A. Damascelli, K.M. Shen, F. Ronning, D.L. Feng, H. Eisaki, Z.-X. Shen, P.K. Mang, N. Kaneko, M. Greven, Y. Onose, Y. Taguchi, and Y. Tokura, cond-mat/0201119.
8. X.J. Zhou, P. Bogdanov, S.A. Kellar, T. Noda, H. Eisaki, S. Uchida, Z. Hussain, and Z.-X. Shen, Science **286**, 268 (1999).
9. A. Ino, C. Kim, M. Nakamura, T. Yoshida, T. Mizokawa, A. Fujimori, Z.-X. Shen, T. Kakeshita, H. Eisaki, and S. Uchida, Phys. Rev. B 65, 094504 (2002).
10. R.S. Markiewicz, Phys. Rev. B**62**, 1252 (2000).
11. C. Kusko, R.S. Markiewicz, M. Lindroos, and A. Bansil, cond-mat/0201117.
12. N.D. Mathur, *et al.*, Nature **394**, 39 (1998).

13. J.L. Tallon, J.W. Loram, G.V.M. Williams, J.R. Cooper, I.R. Fisher, J.D. Johnson, M.P. Staines, and C. Bernhard, Phys. Stat. Sol. b**215**, 531 (1999).
14. N. D. Mermin and H. Wagner, Phys. Rev. Lett. **17**, 1133 (1966).
15. R.S. Markiewicz, unpublished.
16. N. Nagaosa, "Quantum Field Theory in Strongly Correlated Electronic Systems", (Springer, Berlin, 1999), Ch. 3.
17. P. Monthoux and D. Pines, Phys. Rev. B**47**, 6069 (1993); B.P. Stojković and D. Pines, Phys. Rev. B**55**, 8576 (1997).
18. A. Abanov, A.V. Chubukov, and J. Schmalian, cond-mat/0107421, to be published, Adv. Phys.
19. H.Q. Ding, Phys. Rev. Lett. **68**, 1927 (1992).
20. S. Chakravarty, B.I. Halperin, and D.R. Nelson, Phys. Rev. Lett. **60**, 1057 (1988).
21. F. Onufrieva, P. Pfeuty, and M. Kiselev, Phys. Rev. Lett. **82**, 2370 (1999); F. Onufrieva and P. Pfeuty, Phys. Rev. B**61**, 799 (2000).
22. J.C. Campuzano, H. Ding, M.R. Norman, H.M. Fretwell, M. Randeria, A. Kaminski, J. Mesot, T. Takeuchi, T. Sato, T. Yokoya, T. Takahashi, T. Mochiku, K. Kadowaki, P. Guptasarma, D.G. Hinks, Z. Konstantinovic, Z.Z. Li, and H. Raffy, Phys. Rev. Lett. **83**, 3709 (1999).
23. J.L. Tallon, J.R. Cooper, P.S.I.P.N. de Silva, G.V.M. Williams, and J.W. Loram, Phys. Rev. Lett. **75**, 4114 (1995).
24. N. Miyakawa, P. Guptasarma, J.F. Zasadzinski, D.G. Hinks, and K.E. Gray, Phys. Rev. Lett. **80**, 157 (1998).
25. N. Ichikawa, S. Uchida, J.M. Tranquada, T. Niemöller, P.M. Gehring, S.-H. Lee, and J.R. Schneider, Phys. Rev. Lett. **85**, 1738 (2000).
26. J.R. Schrieffer, X.G. Wen, and S.C. Zhang, Phys. Rev. B**39**, 11663 (1989).
27. J.M. Tranquada, B.J. Sternlieb, J.D. Axe, Y. Nakamura, and S. Uchida, Nature **375**, 561 (1995); J.M. Tranquada, J.D. Axe, N. Ichikawa, A.R. Moodenbaugh, Y. Nakamura, and S. Uchida, Phys. Rev. Lett. **78**, 338 (1997).
28. K.M. Lang, V. Madhavan, J.E. Hoffman, E.W. Hudson, H. Eisaki, S. Uchida, and J.C. Davis, Nature **415**, 412 (2002); S.H. Pan, J.P. O'Neal, R.L. Badzey, C. Chamon, H. Ding, J.R. Engelbrecht, Z. Wang, H. Eisaki, S. Uchida, A.K. Gupta, et. al., Nature; **413**, 282 (2001).
29. N. Harima, J. Matsuno, A. Fujimori, Y. Onose, Y. Taguchi, and Y. Tokura, cond-mat/0103519.
30. M. Ambai, Y. Kobayashi, S. Iikubo, and M. Sato, J. Phys. Soc. Jpn. **71**, 538 (2002).
31. R.S. Markiewicz, cond-mat/0108075, to be published as a Comment in Phys. Rev. Lett.
32. R.S. Markiewicz and C. Kusko, cond-mat/0102452, to be published, Phys. Rev. B.
33. E.J. Singley, D.N. Basov, K. Kurahashi, T. Uefuji, and K. Yamada, Phys. Rev. B**64**, 224503 (2001).
34. A. Singh, Z. Tešanović, H. Tang, G. Xiao, C.L. Chien, and J.C. Walker, Phys. Rev. Lett. **64**, 2571 (1990).
35. R.J. Birgeneau, H.J. Guggenheim, and G. Shirane, Phys. Rev. B**1**, 2211 (1970).

ELECTRONIC MOLECULES AND STRIPES IN METALS AND INSULATORS

F.V.KUSMARTSEV
Department of Physics, Loughborough University LE11 3TU, England
E-mail: F.Kusmartsev@lboro.ac.uk

We demonstrate that electronic clusters in the form of molecules, stripes or droplets may arise in various media. In liquid Helium or metal-ammonia solutions they have a spherically symmetric form of a bubble. In narrow band ionic insulators, they may form strings or stripes which may arise due to Fröhlich electron-phonon interaction alone. If it is an insulating state then the system consists of isolating clusters. If it is a metallic state the system is in an inhomogeneous state consisting of a liquid of strings embedded into Fermi see. The conditions for the string existence and the string formation, the string length and the number of particles self-trapped into the string depend on the electron(hole) conduction bandwidth and on the ratio of high frequency and static dielectric constants. Such a state may arise in a wide range of physically interesting parameters and we show that such inhomogeneous insulating and conducting states with electron strings relevant to oxide materials, like HTSC.

1 Introduction

The formation of electronic clusters due to a deformation of the media has been known for a long time. For example, a single and multi-electron bubbles arise in liquid helium and metal ammonia solutions. In the Motts model for a polaron applied to metal-ammonia solutions a single electron is self-trapped by a vacuum bubble. A similar bubble is arising in liquid helium. The bubble may trap more than one electron, which will effectively form an electronic cluster which is trapped inside vacuum bubble. Such a phenomenon has been studied well theoretically and experimentally. The similar phenomena may also arise in solids, where, however, their origination is less transparent. Recently we have introduced the notion of electronic molecules (e-molecules) [1,2,3], which may arise in solids with strong or intermediate electron-phonon interactions. The e-molecule consists of a few electrons bound by electron-phonon or by electron-electron interactions. The e-molecule is a generalization of the notion of a polaron[4,5,6] to a many-body case adopted to solids with narrow bands. We have shown that nearly all types of the electron phonon interaction will lead to the creation of e-molecules, which may be created either in a ground state or in a metastable state and have in many cases a linear shape. Due to such a linear shape they have been named as strings[1,2,3]. At present there is a growing body of experimental evidence[7,8,9,10] indicating the existence of complex inhomogeneous micro- and mesoscopic structures created in oxide materials.

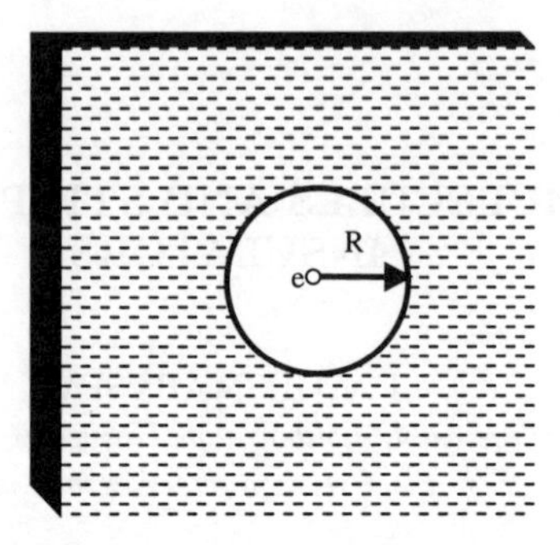

Figure 1: The single electron bubble in liquid Helium.

Such structures are typically connected to *static or dynamic stripe phases*, which were independently proposed by many researchers in their seminal papers[6,11-20]. More recently the importance of lattice distortions in the creation of these stripe, like structures[9,8,10] has been emphasized. All these static and dynamic structures observed in HTSC[15-9,7] may have a natural explanation in the framework of the general notion of e-molecules. Recent experiments have also discovered a huge influence of isotope effect on the critical temperature of the stripe ordering[9] and strong lattice fluctuations in YBCO[8] which may be associated with the dynamics of the strings or e-molecules. With the isotope changes[9] the structure of individual strings is changed (for example, the strings become shorter) and, therefore, the critical temperature of the stripe ordering must change. This indicates that the electronic molecules and, in particular, strings, are a pertinent notion for oxides and, possibly, also for other solids if experiments will finally clearly verify them.

2 Electron Bubbles or a Polaron or E-molecules in Metal-Ammonia Solutions

An electron ejected into the liquid Helium creates a special state, so-called an electron bubble. The similar electron bubble is arising in metal-ammonia solutions and named as a polaron, after Mott. The matter is that the work function of this electron from the bulk helium into the vacuum is negative and approximately is equal to about $1eV$. Therefore to gain this energy the electron expels the helium atoms around and creates a vacuum bubble of the radius R in which this electron is self-trapped. As a first approximation the energy of this self-trapped electron may be estimated as the following sum:

$$E = E_{el} + E_{surface} + W$$

where the value E_{el} the work function of the electron into the vacuum plus the kinetic energy of the electron trapped into the the bubble. The value $E_{surface}$ is the surface energy of the bubble and the value W is a work performed to create a bubble against the pressure p: $W = \int pdV$.

If we count the electron energy from the vacuum and assume that the bubble has an infinite high potential wells then the value E_{el} is equal to

$$E_{el} = \frac{h^2}{8mR^2},$$

where m is an electron mass and h is a Planck constant. For the spherical bubble the surface energy has a standard form as $E_{surface} = \sigma S$ where S is a surface of the bubble and σ is a surface tension, i.e.

$$E_{surface} = 4\pi R^2 \sigma.$$

Finally the work against the external pressure needed to create the bubble is $W = pV$, i.e.

$$W = \frac{4}{3}\pi R^3 p$$

If the external pressure is zero the lowest energy is obtained for the bubble with the radius found from the equation

$$\frac{\partial E}{\partial R} = 0 = -\frac{h^2}{4mR^3} + 8\pi R\sigma.$$

The solution of this equation gives the radius of the spherical bubble containing an electron in the ground state, R_0 as:

$$R_0 = \left(\frac{h^2}{32\pi m\sigma}\right)^{1/4},$$

At this moment there is an issue which is intensively discussed in the literature, namely, what is a structure of the excited state of such an electron bubble?

2.1 *Electrino => Electron Cold Fission or Confusion?*

Maris recently proposed that an usual electron may be split into two particles named as electrino (see, for example, Refs[21,22,23]). Such an electron fission will violate main fundamental principles of quantum mechanics and quantum field theory, which claim that the fraction of an electron, like for example, quarks, can not exist. Maris claims that such an electron fission may arise

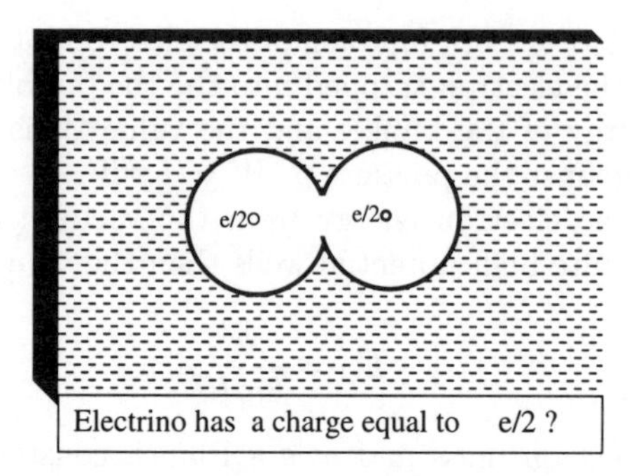

Figure 2: The hypotetical proccess of the electron fission, which probably never arises

for electrons in liquid helium. Inside the helium a spherical vacuum bubble is created around each electron. This bubble arises because the same electron in the vacuum has a lower energy than inside the Helium. The energy gain is of the order of 1eV. Therefore the electron in Helium likes to have a spherical vacuum bubble around and this is the lowest ground state for the electron. This phenomenon is known for four decades.

The Maris conclusion about an existence of electron fission is based on the suggestion that under a light excitation the bubble may get an equilibrium "dumb-bell" shape. With the increase of the pressure the waist in the dumb-bell may vanish and then two nearly spherical small bubbles, each, trapping one half of the electron, arises. Maris has named these hypotetical particles as electrinos.

Is this really true that the main laws of physics are violated in liquid Helium and we have a novel particle or a quasi-particle: an electrino?. It is very difficult to argue against such a cold fission. However, let us note that the dumb-bell bubble may be viewed as a double well potential. In the double well in a contrast with a single well an electron energy level is split into two levels associated with symmetrical and anti-symmetrical states. According to a main law of quantum mechanics these states must be orthogonal each other and they do. But the symmetrical state of the dumb-bell bubble associated with the lowest energy is unstable. As a result of such instability a spherical bubble having a lowest ground state energy is created. The other anti-symmetrical dumb-bell state remains metastable. It seems that in such a picture the electron fission is possible. However the missing point here is an orthogonality of the states associated with the spherical and the dumb-bell bubbles. So, if before the instability developed the symmetric and the asymmetric states in the dumb-bell bubble were orthogonal each other, then after the creation of

Figure 3: Bi-Polaron has a charge equal to 2e while E-molecule has a charge equal to Me.

the spherical bubble these states are already not orthogonal each other. This means that that the dumb-bell bubble does not correspond to an excited equilibrium state of the spherical electron bubble. Therefore, the electron state in the dumb-bell bubble does not exist at all since it is not orthogonal to the ground state associated with the spherical bubble and therefore unstable. The situation is very general and already has been intensively discussed in the past for other type of self-trapped electrons by Pekar[4] and Rashba[5], pioneers of polaron theory. The conclusion was that there always arise only one self-trapped state of the lowest energy while all other existing states are associated with the free band electrons. In the considered case the self-trapped state is the spherical bubble. Therefore the dumb-bell bubble does not exist, can not be split into two parts and two electrinos bubbles can not be created. Thus, the main laws of the quantum mechanics and the quantum field theory do not violate. The process opposite to the discussed electron cold fission is a creation of the multi-electron bubbles. The simplest of them is a bound state of two-particles known as a bipolaron, although typically these bubbles may trap many electrons (see, Fig.3.) However in both cases for the electrino and for the bi-polaron it is very difficult to find direct experimental evidences indicating on their existence, while single and multi-electron bubbles have been clearly identified.

3 E-molecules in ionic solids

The electronic molecule is a many-particle generalization of a conventional polarons. The electron strings or linear molecules, may be created both by a long- and a short-range electron-phonon interaction[1,2] and they can be highly-conducting or completely insulating defects. In this section we show that the Fröhlich interaction combined with direct Coulomb repulsion does lead to a charge segregation like strings in doped narrow band insulators, both in adiabatic and in nonadiabatic regimes confirming the main conclusions of the original papers about a creation of strings[1,2].

To investigate the creation of the string we have introduced a new approach based on the finding of exact solutions for discrete Hartree-Fock equations (see, for example,Ref.[24]) and have found . that the creation of multi-particle self-trapped states is possible. Similar states originated due to a rotation of mobile molecules in liquid polymers have been proposed by Grigorov et al.[32]. Our previous estimations indicated that for typical physical parameters, for example, for oxide superconductors, the strings have small length, of the order of a few of the interatomic lattice spacings, and constitute of a small number of the trapped particles. Therefore, to describe a structure of these electronic molecules as well as to find a precise criterion for their existence we have to perform a proper consideration of the lattice discreteness. Doing this, without going to simplifications of the continuum approximations, we find the criterion for the formation of both conducting and insulating strings in ionic insulator taking into account the finite bandwidth and treating the kinetic energy of electrons on equal footing with their potential energy. Thus, in the present paper we investigate a formation of electronic strings in ionic solids using the approach presented in the papers[1,2]. We consider a very general Hamiltonian for spinless fermions interacting with phonons and with each other via long-range Coulomb forces and located on a $d-$ dimensional hypercubic lattice:

$$H = -t \sum_{<i,j>} a_i^\dagger a_j \quad + \sum_{q,i} \omega(q) n_i [u_i(q) b_q + h.c.] + \\ \sum_q \omega(q) b_q^\dagger b_q + \sum_{i<j} V(i-j) n_i n_j . \tag{1}$$

there t is the electron hopping-integral, the operator $a_i^\dagger (a_i)$ creates (destroys) a fermion at a lattice site i , n_i is the occupation number operator $a_i^+ a_i$ and the operator $b_q^\dagger (b_q)$ is an operator of the creation (destruction) of a phonon. The summations in eq.(1) extend over the lattice sites i and, as indicated by $< i,j >$, over the associated nearest sites j. The matrix element of the electron-phonon interaction is equal to

$$u_n(q) = \frac{\gamma(q) \exp{(iqn)}}{\sqrt{2N}} \tag{2}$$

The function $\gamma(q)$ and the phonon dispersion relation $\omega(q)$ are different for different types of electron-phonon interactions. The function $V(i-j) = e^2/|i-j|a$ is a two-particle potential of the long-range Coulomb interaction[1,2].

To describe a string with M spinless fermions localised in the potential well consisting of N lattice sites, we employ a Hartree-Fock many-body wave function, $\Psi(1,2,...,M)$, which has the form of a Slater determinant (see, also in Refs[1,2])

$$\Psi(1,2,...,M) = \frac{1}{\sqrt{M!}} \det || \psi_i(k_j) || \quad (3)$$

consisting of normalised single particle wave functions:

$$\psi_{m_x}(k_j) = \begin{cases} \frac{1}{\sqrt{N}} \exp(ik_j m_x) & \text{if} \quad 1 \leq m_x \leq N \\ 0, & \text{otherwise.} \end{cases} \quad (4)$$

Each of these wave functions describes the electron(hole) trapped by N neighboring sites (string potential well) with equal probability, $1/N$. If the string is oriented in the **x** direction and is located on the sites $m_x = 1,...,N$ the particle quasi-momentum k_j is determined by boundary conditions, which is naturally taken as open one[24]. The employed Hartree-Fock many body-wave function was obtained originally as an exact wave function for M free fermions self-trapped by a string potential well[24]. Outside the self-trapped potential well this wave function is vanishing. This wave function is also an exact solution of the Hartree Fock equations for fermions trapped into a string configuration.

For electrons (or holes) interacting with polar phonons, i.e. with longitudinal optical phonons with frequency ω_q the constant of the electron- phonon interaction $\hbar\omega(q)\gamma^2(q) = 4\pi e^2/(q^2\epsilon^*)$ [4] with $1/\epsilon^* = 1/\epsilon_\infty - 1/\epsilon_0$, where ϵ_∞ is a high frequency– and ϵ_0 is a static– dielectric constants. The value of total energy may be estimated with the use of the Pekar functional $J = T + V + V_c$ which includes the conventional kinetic and potential energy of fermions T and V, respectively, together with the new contribution, V_c, coming from the long-range Coulomb interaction between fermions(see, also for example, in the Refs[1,2]).

With the aid of the Hartree-Fock many-body wave function of the M self-trapped particles $\Psi(1,2,...,M)$ (see, eq.(3)) we have exactly calculated all these contributions T, V and V_c[1,2]. The exact expression for kinetic energy is

$$T = 2dt - \frac{2t\ (1-1/N)\ \sin(\pi\, M/N)}{M \sin(\frac{\pi}{N})} \quad (5)$$

This universal form of kinetic energy is valid both for odd and for even number of particles trapped into the string, contrary to an erroneous statement in Ref.[25]. To get this expression the boundary conditions have been properly adjusted to both these cases of the odd and even number of particles to have the vanishing total momentum of all electrons. The matter is that to keep the total momentum of fermions trapped into the string equal to zero we have to use different boundary conditions for odd and for even number of particles, ie periodic and antiperiodic ones, respectively. The potential energy associated with the electron phonon interaction has a conventional ala Pekar form[1,2]:

$$V = E_p M I_N \tag{6}$$

where we have introduced the notations, $E_p = E_c(1 - \epsilon_\infty/\epsilon_0)$ and the value $E_c = e^2/(a\epsilon_\infty)$ with a the interatomic distance. The integral I_N has been defined in Ref.[2] as

$$I_N = \int_{-\pi}^{\pi}\int_{-\pi}^{\pi}\int_{-\pi}^{\pi} \frac{dxdydzG(x,y,z)}{(2\pi)^3}\left(\frac{\sin(Nx/2)}{N\sin\frac{x}{2}}\right)^2 \tag{7}$$

where $G^{-1}(x,y,z) = 3 - \cos x - \cos y - \cos z$ is a 3D lattice Greens function. With the increase of N the value of I_N decreases as $I_N \approx \frac{A}{N^\alpha}$, where A and α are some near constant parameters. The dependence of A and α on N is extremely weak. So for $10 \leq N \leq 2.0\ 10^3$ the value I_N can be well extrapolated with $A \approx 0.9743$ and $\alpha \approx 0.85$. For the range of the numbers for the string length $N > 10$ the continuum approximation works very well, while in the limit $N \to \infty$ the value $I_N \sim lnN/N$[25,2]. Although for realistic physical situations associated with small number N this precise asymptotic limit is not relevant.

To estimate the contribution from the Coulomb repulsion between particles trapped into the string, V_c, first we have to calculate the discrete density matrix. Using the many-body wave function which is a Slater determinant built up of the single particle wave functions for the electrons trapped into the string, eqs(3,4) (see, Refs[1,2],for a detail) we obtain that this density matrix has the form

$$\rho_M(n,m) = \frac{M}{(M-1)N^2}\left(1 - \frac{\sin^2(\pi(n-m)M/N)}{M^2\sin^2(\pi(n-m)/N)}\right). \tag{8}$$

This matrix is, of course, properly normalised $\sum_{n,m=1}^{N} \rho_{M,N}(n,m) = 1$. With the use of this density matrix the Coulomb energy may be expressed via the two-body interaction potential $V(n-m)$ as

$$V_c = \frac{M(M-1)}{2}\sum_{n,m} V(n-m)\rho_{M,N}(n,m) \tag{9}$$

where the Coulomb forces between particles are described by a conventional two particle potential of the form

$$V(n-m) = \frac{e^2}{\epsilon_\infty a|n-m|} \tag{10}$$

where ϵ_∞ is a high frequency dielectric constant. Then after the use of the two-particle density matrix we have that

$$V_c = \frac{M^2}{2N^2}\frac{e^2}{\epsilon_\infty a}\sum_{n,m}\left(1 - \frac{\sin^2(\pi(n-m)M/N)}{M^2\sin^2(\pi(n-m)/N)}\right)\left(\frac{1}{|n-m|}\right) \tag{11}$$

The state associated with the electronic molecule is determined by a minimization of the total energy J with respect to the length of the string N at the fixed value of the number of particles M trapped into the string. Each such a state corresponds to a local minimum of the total energy and therefore is a stable. Some string may have a lowest energy per particle, then this string is in a ground state. The minima associated with the different electronic molecules may be located higher or lower than a position of the conduction band. However because of the local stability, such states will be always realised at some physical conditions, although an occupation number of these states may be not as large.

A simplest way to analyse the existence and physical parameters of these electronic molecules is to use a continuum approximation of the obtained exact expressions. However such a continuum approximation has used in some previous papers and has a very limited range of application when the number of particles trapped is small. But in most physically interesting situations the strings have small number of particles which are trapped on the small number of sites $M \sim N$. The value of M is of the order of 2-10 particles. Therefore, in this case the discreteness of the lattice must be properly taken into account. Using the discrete functional J we have calculated the string energy for the physical parameters $\epsilon_\infty = 5$ and $\epsilon_0 = 30$ and for the lattice parameter $a = 3.8A$. The value of the hopping integral was chosen as $t = .5eV$. The result is presented in Fig.4.

The parabolic curve fitted to a first group of points as shown in Fig.4 corresponds to a separate electronic molecule having a fixed different number of particles. Other groups of the points may be fitted by other parabolic curves associated with different e-molecules. All of these molecules are in metastable states and therefore at zero temperatures will not be realised unless the Fermi energy will be located at the string energy or above. In fact, in this case the lowest energy string will supply a fixed localized level which may pin the Fermi

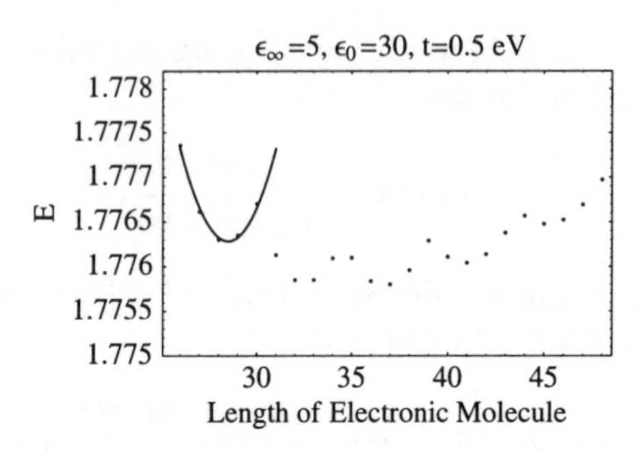

Figure 4: The energy of strings for the physical parameters $\epsilon_\infty = 5$ and $\epsilon_0 = 30$, the lattice parameter $a = 3.8A$, the value of the hopping integral is $t = .5eV$. The parabolic curve is drawn for an eye guide. One sees that there are many group of points which may fit to analogous parabolic curves, each of which corresponds to a string with a fixed number of particles.

energy. With doping the electrons (or holes) will be filling this string energy level (see, Fig.4). The similar situation has been originally proposed by Gorkov and Sokol[26].

The position of these minima may be located not only above the minimum of the conduction band but also below. For example, in the above considered example this happens when the bandwidth of the conduction band will be smaller than $.5eV$. However, when the value of high frequency dielectric constant will be changed, the strings may have energy lower than the bottom of conduction band in wide range of the physical parameters. For example, when $\epsilon_\infty = 2$, the strings energy levels will be located below the conduction band (see, Fig.5) already when the bandwidth, $W = 6t = 1.5eV$.

Due to Fröhlich electron-phonon interaction the string energy may be even smaller than the energy of the Pekar polaron[4]. Such Pekar polarons are usually playing a role of itinerant current carriers in ionic solids. Since the energy of the Pekar polarons is located below the minimum of the conduction band associated with the free electrons, the strings having the lowest energy is in a ground state. The binding energy of the Pekar polaron found originally by Pekar[4], with the use of an effective mass method this energy may be expressed in the form:

$$E_{PEKAR} = -0.0154 \frac{me^4}{2ta^2\epsilon^{*2}}. \tag{12}$$

Now let us compare an energy of the optimal string (that is the string or a linear molecule, which has the lowest energy per particle among all linear electronic molecules) with the binding energy of the Pekar polaron when the

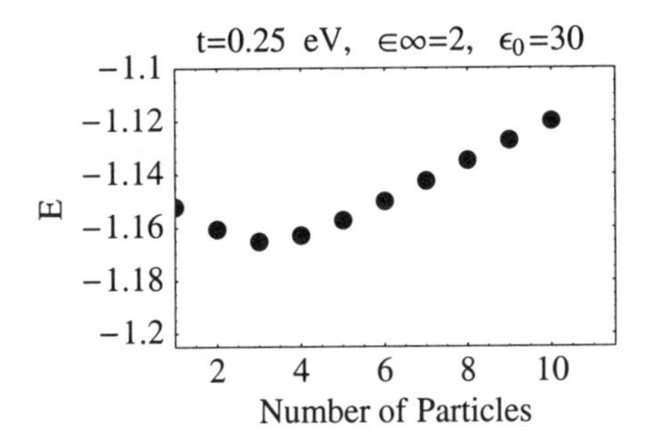

Figure 5: The dependence of the string energy on the particle number. These energy levels for strings having 2,3,4 and 5 number of particles trapped are presented by points. All these energy levels are located below the conduction band, which bottom is located at zero energy.

value of the bandwidth $W = 6t$ varies and all other parameters (dielectric constants and an interatomic spacing) are fixed. We investigate and compare the dependencies of the optimal string energy and the Pekar polaron energy on the bandwidth W or on the hopping integral t. Our investigation shows that if the ratio of dielectric constant is large enough (ie it is larger than some critical value) the optimal string energy per particle becomes smaller than the polaron energy if the bandwidth is smaller some other critical value,ie $t < t_c$. For an illustration we present this energy dependency on t in Fig.6 where the values the dielectric constants are taken to be $\epsilon_0 = 30$ and $\epsilon_\infty = 2$.

From the Fig.6 we see that if the bandwidth or the hopping integral t is smaller than some critical value $t < t_c = .43eV$ the optimal string is in a ground state, indeed. This conclusion is opposite to erroneous conclusion obtained in Ref.[25], based on the estimation made for the physical parameters associated with strings in metastable states. The value of t_c depends also on the ratio of the dielectric constants, ie, $t_c(\epsilon_0/\epsilon_\infty)$, and increases when the ratio of these dielectric constants $\epsilon_0/\epsilon_\infty$ increases. For the chosen values of dielectric constants the strings arise in ionic solids when the bandwidth is of the order of 1-2 eV or smaller. However with the next decrease of the bandwidth the string solutions disappear and the strings collapse into a small polaron state to which the Pekar polaron state is transformed, too.

There arises an abrupt transition from the string state to the small polaron state. With decreasing bandwidth the length of the string and the number of particle trapped changes (decreases) abruptly because each string corresponds to a local minimum of the total energy and separate from other state by a barrier. Therefore, any of these transitions must go over the barrier. With

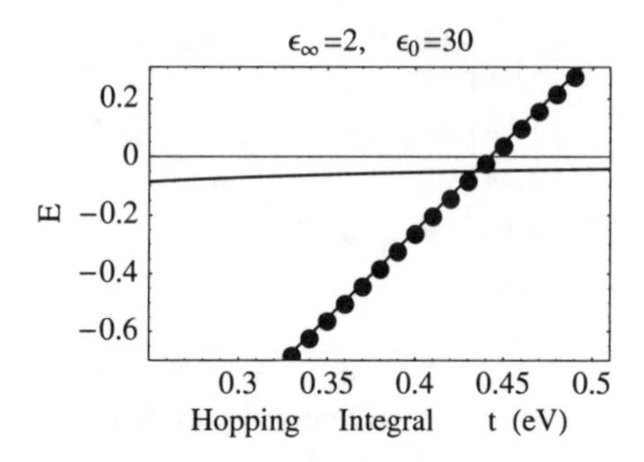

Figure 6: We present the optimal string energy dependence on the value of the hopping integral t,both measured in eV, by points. For a comparison, the Pekar and small polaron energies are presented by solid curves. The values of the dielectric constants are taken to be equal to $\epsilon_0 = 30$ and $\epsilon_\infty = 2$.

decreasing bandwidth the number of particles trapped becomes smaller and smaller. For example, with decreasing bandwidth, first, a four particle string arises, then - a three-particle string(tri-polaron), then - a bipolaron and finally a last, final transition to a small polaron. It is important to note that for the wide bands the string corresponds to a metastable state, has a very large length and is trapping many particles. Here the continuum approximation[2,29] for the number of particle trapped and for the length of the string, gives reliable results. On the other hand for narrow bands, the string length is small and inside the string there are only a few particles trapped. At some conditions such a string state may be a ground state of the system.

Of course there are many other types of the phonons in ionic solids. If we taken into account a contribution from other type of the electron-phonon interaction into a formation of electronic molecules we obtain that the conditions for the string formation are strongly improved. For example, for a string to be in a ground state the range of values for the bandwidth and for the ratio of dielectric constants becomes broader. The reason is that any contribution from other type of electron-phonon interaction is mostly improving the potential well trapping the particles while the contribution from Coulomb repulsion remains effectively the same.

Thus, we arrive at the conclusion that in oxide compounds with ionic bonding the formation of electron strings created by a polarization potential only is possible. The string length is typically much larger than the number of self-trapped particles, which is determined by the dielectric constants of the solid and the value of the bandwidth (see, also Refs[27,28,29,30,31]).

In general the strings may correspond to either a ground state or a metastable state and arise typically when the bandwidth takes intermediate values. With the increase of the bandwidth the ground state associated with these strings transforms into a metastable state while the string length and the string degeneracy increase. Depending on this there may arise different physical situations, like, for example, phases of static and dynamic stripes, of superconducting and insulating states.

In summary, we have shown that strings may arise in ionic solids due to an interaction with Fröhlich phonons alone. Any other type of the electron-phonon interaction taken into account strongly improves and increases the range of the physical parameters at which these strings arise in a ground state. We believe that such electronic molecules are pertinent notion for oxides and possibly also for other materials where may arise an interplay between Coulomb forces and electron-phonon interactions.

Acknowledgments

I am very grateful to S. Kivelson, D. Edwards, G. Gehring, V. Emery, E.I. Rashba, Danya Khomskii, Klim Kugel, Ania Myasnikova and Mikko Saarela for illuminating discussions and D.M. Eagles on indicating the existence of the analogous polaron instability and the consequential formation of electron strings in liquids of polymers[32]. The work has been supported in part by Isaac Newton Institute, University Of Cambridge.

References

1. F.V. Kusmartsev, J. de Physique IV,**9**, Pr10-321, (1999)
2. F.V. Kusmartsev, Phys.Rev.Lett. **84**, 530 (2000); Phys.Rev.Lett. **84**, 5036 (2000)
3. F.V. Kusmartsev, Physica B284-288 (2000) 1422; EuroPhys. Lett. **54** (2001) 786; ibid **57** (2002) 557
4. S.I. Pekar, Untersuchungen über die Elektronentheorie Kristalle, Akademie Verlag, Berlin, 1954.
5. E.I. Rashba, in: Excitons, ed. by E.I.Rashba and M.D. Sturge, North-Holland (Amsterdam) 1982, p.543.
6. E.L. Nagaev, Sov. Jour.- JETP Lett., **16**, 558 (1972); V.A. Kaschin and E.L. Nagaev, Zh. Eks. Teor. Fiz., **66**, 2105 (1974)
7. H.A.Mook, Pengcheng Dai, F. Dogan, R.D. Hunt, Nature, **404** 729 (2000)
8. R.P. Sharma et al, Nature, **404**, 736 (2000)
9. A. Lanzara, G. Zhao, N. L. Saini, A. Bianconi, K. Conder, H. Keller and K. A. Müller, J. Phys. Cond. Matt. **11**, L541 (1999)

10. M.Uehara, S. Mori, C.H. Chen and S. W. Cheong, Nature,**399** 560 (1999) ; S. Mori, S. Chen et al, **392**, 473 (1999)
11. J.R.Zaanen and O. Gunnarson, Phys. Rev. **B40**, 7391 (1989)
12. U. Löw, V. J. Emery, K. Fabricius and S.A. Kivelson, Phys. Rev. Lett. **72**, 1918 (1994)
13. V.J. Emery, and S.A. Kivelson, Nature (London) **374**, 434, (1995); V.J. Emery, S.A. Kivelson, and O. Zachar, Phys. Rev, **B56**, 6120, (1997).
14. V.J. Emery, S. Kivelson and H.Q. Lin, Physica **B163**, 306, (1990); Phys. Rev. Lett. **64**, 475, (1990)
15. A. Bianconi, Phys. Rev. **B54**, 12018 (1996); M.v. Zimmermann et al, Eur.Phys. Lett. **41**, 629 (1998).
16. T.R.Thursten et al, Phys. Rev. **B40**, 4585 (1989).
17. J.M. Tranquada, Nature (London) **375**, 561, (1995).
18. A. Bianconi et al, Phys. Rev. Lett. **76**, 3412 (1996); and see Refs therein.
(1995);
19. H.A. Mook, P. C . Dai, S.M. Hayden, G. Aeppli, T. G. Perring and F. Dogan, Nature, **395** 580 (1998)
20. N.L.Saini, J.Avila, A.Bianconi, A.Lanzara, M.C.Asensio, S.Tajima, G.D.Gu and N.Koshizuka, Phys. Rev. Lett. **79**, 3467 (1997)
21. Maris H.J., J. Low Temp. Phys., **120**, 173 (2000)
22. Chown M., New Scientist, **N2260**, 25 (2000)
23. Physics Today, **53**, 9 (2000)
24. H.S. Dhillon, F.V. Kusmartsev and K. Kuerten, Phys. Rev. **B60**, 6208 (1999) and J. Nonlinear Math. Physics, **8**, 38 (2001).
25. A.A. Aleksandrov and V.A. Kabanov, Pisma Zh. Eksp. Teor. Fiz. **72**, 825 (2000)
26. L.P.Gorkov and A.B.Sokol, Pisma Zh. Eksp. Teor. Fiz. **46**, 333 (1987)
27. F.V. Kusmartsev, D. Di Castro, G. and A.Bianconi, Phys.Lett., **A275** (2000) 118.
28. X.G. Zheng et al, Phys.Rev. Lett. **85**, 5170 (2000)
29. F.V. Kusmartsev, Int. J. Modern Phys., **14**, (2000) 3530;
30. T. Egami, in book:"Physics of Local Lattice Distortion" ed. by H. Oyanagi and A. Bianconi, AIP, **V554** (2001) 38
31. V.J. Emery, S.A. Kivelson and J.M. Tranquada, Proc. Natl. Acad. Sci. USA **96**, 8814 (1999)
32. L.N. Grigorov, Makromol. Chem., Macromol. Symp. **37**, 159 (1990).; L.N. Grigorov, V.M. Andreev, and S.G. Smirnova, Makromol. Chem., Macromol. Symp. **37**, 177 (1990); L.N. Grigorov, Pis'ma Zh. Tekh. Fiz. **17** (5) (1991) 45. [Sov. Tech. Phys. Lett. **17**, 368 (1991).

COMPOSITE TEXTURED POLARONS IN COMPLEX ELECTRONIC OXIDES

S. R. SHENOY
Abdus Salam International Centre for Theoretical Physics
Trieste 34100, Italy
E-mail: shenoy@ictp.trieste.it

T. LOOKMAN, A. SAXENA AND A. R. BISHOP
Theoretical Division,
Los Alamos National Lab,
Los Alamos, NM 87544, USA
E-mail: txl@lanl.gov

We study a Ginzburg-Landau model with coupled strain, charge and magnetization variables. 'Intrinsic inhomogeneities' appear naturally, and are understood as composite and textured multipolaron responses of a nonlinear lattice, to random doping perturbations, with a constraint that lattice integrity is maintained.

1. Introduction

High-resolution microscopies in complex electronic oxides have revealed an unsuspected non-uniform nanoworld of modulated strain, magnetization, and charge density, or orientation of charge orbitals[1,2,3,4,5,6,7]. Materials include martensites, high-temperature superconducting (HTS) cuprates, colossal magnetoresistance (CMR) manganites, relaxor ferroelectrics and other compounds. The 'textured', or sign-varying spatial patterns, form meandering ribbons, stripes or droplets over nano- or meso-scales, that vary with temperature, and with uniform external stress, magnetic field or voltage[8,9,11,12,13]. The ubiquitous nature of these spatial variations suggests a search for a generic explanation, (that will of course depend in its details, on material parameters). A common element seems to be the coupling of charge, spin and *strain*, with the intriguing composite textures occuring below (or just above) a structural phase transition that in the most important cases is *ferroelastic.*

The ferroelastic transition is displacive in nature, with atoms shifted by less than a lattice parameter, and with components of the strain tensor as the order parameter. Below the transition there are *displacively related competing ground state structures* differing by the signs or values of the strain order parameter(s). Ferroelastics like FePd alloys (that are shape memory materials) have a face-centred cubic to face-centred tetragonal transition, while cuprates have a tetragonal-to-orthorhombic transition[15]. Shape-memory alloys, cuprates, manganites, titanates and other materials[16,17], often show characteristic texturing of strain variables around the structural transition, with e.g. nanoscale criss-cross 'tweed' patterns above transition; and parallel elastic domain walls, or 'twins' below it.

Both HTS cuprates and CMR manganites show evidence of coupling between charge, spin, and these strain degrees of freedom. In cuprates, the structural transition temperature, as well as the Néel antiferromagnetic temperature fall steeply with doping[1,14]. The Néel temperature, the superconducting transition temperature T_c, and the pseudogap temperature T^* also show large isotope effects[18], that can be site-specific. Microstructure memory effects occur in cuprate needle twins, under application of a magnetic field[13]. In manganites, the Curie ferromagnetic temperature shows a giant isotope effect[19], and in addition to colossal magnetoresistance, there can be 'colossal stressoresistance' (CSR), under application of hydrostatic pressure. X-ray and neutron diffuse scattering in manganites reveals four-lobe Fourier space images attributed to polarons, that vary with magnetic field[12], as do diffuse 'cloud-like' regions of high and low conductivity in STM or TEM coordinate space pictures[9,10].

A Ginzburg-Landau (GL) description has been useful in work on structural texturing, with free energy as symmetry-allowed powers of the strain components[20,21,22,23,24,25]. In general, spatially varying patterns emerge from a competition between short- and long-range forces[26] and in the structural case, it is the constraint[27] of *elastic compatibility* that induces the crucial anisotropic, long-range (ALR) force. Tweed and twins appear in strain models, with this ALR force acting. While there is much theoretical work on microscopic models[28,29,30,31,32,33,34,35,37], it does not usually include coupling to a nonlinear lattice. As a conceptual guide to further microscopic investigations, it is natural to extend our GL strain model[22] to include symmetry-allowed couplings to charge and magnetization variables, to see what new textures are induced by the charge. We thus view the materials as *doped ferroelastics.*

The St. Venant elastic compatibility condition expresses the constraint that the elastic fields have no defects such as dislocations or ionic vacancies, and is a differential equation[27] linking components of the strain tensor $\underline{\underline{E}}$,

$$\vec{\Delta} \times (\vec{\Delta} \times \underline{\underline{E}}(\vec{r}))^{\dagger} = 0, \qquad (1a)$$

or in Fourier space

$$\vec{k} \times \underline{\underline{E}}(\vec{k}) \times \vec{k} = 0, \qquad (1b)$$

and is like the Maxwell equation $\vec{\Delta}.\vec{B} = 0$, where the absence of a source term implies there are no magnetic monopoles. If the strain tensor components are written as derivatives of the displacement vector $\vec{u}$ taken as single-valued, i.e. $E_{\mu\nu} = (1/2)[\Delta_\mu u_\nu + \Delta_\nu u_\mu]$, then Eq. (1) is satisfied as a vector identity, in the same way as introducing a single-valued vector potential through $\vec{B} = \Delta \times \vec{A}$ satisfies the Maxwell equation through 'Div Curl=0'. (We ignore throughout 'geometric nonlinearity' or higher order, and smaller, derivative terms in the strain). Elasticity theory is usually presented in terms of displacement, with strains as derived quantities; the St. Venant condition (identity) is then a useful subsidiary check. This approach is like doing electromagnetism entirely in terms of vector potentials, and then deriving $\vec{B}$. However, it is also instructive to think of the magnetic induction $\vec{B}$ (and similarly, the strain $\underline{\underline{E}}$) as a physical field, when the Maxwell no-monopole condition (and the St. Venant no-defect condition) are elevated to field equations in their own right. Within this viewpoint, as noted by Baus and Lovett, the displacement can then be derived if need be, from this strain[27]. But the focus is on the physical strain variables, that are the natural order parameters in a Landau approach to phase transitions. We will follow this strain-based approach, that has an important consequence: it uncovers the anisotropic long-range (ALR) potentials that arise from the compatibility constraint of lattice integrity, which are left implicit in the displacement picture, and indeed are commonly (and justifiably) ignored in analyses of harmonic-lattice polarons.

A coupling of charge $n(\vec{r})$ to strain $e_1(\vec{r})$ is, in Fourier space, $-\lambda_0 n(\vec{r})e_1(\vec{r}) \rightarrow -\lambda_0 n(\vec{q})\vec{q}.\vec{u}(\vec{q}) = \lambda(\vec{q})n(\vec{q})u(\vec{q})$ where the electron-phonon coupling $-\lambda(\vec{q}) \equiv \lambda_0 \vec{q}.\hat{u}$ absorbs the polarization direction $\hat{u}$. A resulting model hamiltonian

$$H = -\sum_{\vec{q}} \lambda(\vec{q})n(\vec{q})u(\vec{q}) + (1/2)Kq^2 u(\vec{q})^2, \qquad (2)$$

is quite adequate for discussing, e.g., polaron formation in harmonic solids. However, the most interesting effects occur when compatibility meets nonlinearity. In GL models that describe a first-order phase transition, free energies are (in general) sixth powers of the order parameter (OP) strains, while simply harmonic in non-OP strains. The non-OP strains can be eliminated subject to the St. Venant constraint, and their free energy contribution then becomes the ALR potential between order parameters at widely distant points[21,22]. If the non-OP elastic constants are large relative to those of the order parameter, then the system would like to get rid of non-OP strains, and suppress their energetically costly contribution. The diagonal twin textures in deviatoric strain are accompanied by such a Meissner-like expulsion of compressional and shear strains[22]. However such a suppression may not be possible, if under e.g. external stress, non-OP strains are needed to maintain lattice integrity. The constrained system may then seek any other allowed way to lower its energy, even if this means introducing spatial variations of the strains (that are cheap, at the price).

Our central physical idea is quite simple. Suppose, within an otherwise uniform sea of rectangular unit cells, (symmetry-broken to one strain minimum), a single unit-cell is made square by, e.g., a local stress. It is clear that to maintain lattice integrity, the neighbouring unit cells (and their further neighbours) must also deform, with an admixture of non-OP strains, as required. For non-OP energy costs large enough, it may be profitable to generate opposite-sign strain, in *adaptive elastic screening.* For example, in a nonlinear strain model above transition, with strong external local stress, we have found an instructive time evolution (both in underdamped as well as overdamped dynamics): a single-sign 'monopolar' elastic strain is followed by a formation of energy-lowering higher elastic multipoles of opposite sign, in an expanding texture[24]. Since doped charges can act as local stress (or temperature), we expect them to produce some kind of characteristic texture.

Model simulations of the GL model reveal that surprisingly rich textures arise, from three seemingly innocuous assumptions:

(1) The system can undergo a structural transition, with competing ground states.

(2) The free energy is a symmetry-allowed scalar (including couplings between strain, spin, and charge).

(3) The charges deform, but do not rupture, the lattice. We find that charges doped into a nonlinear-strain lattice, below a structural transition and with a magnetizable background, can form an unusual type of composite polaron that we term a *p*olaronic *e*lasto*m*agnetic *t*exture, or '*pemt*on'.

The strain fields have a general 'butterfly' shape, are of both signs, and extend over scores of lattice spacings, with the coupled magnetization also lobed and sign-varying. This is a concrete example of the concept of magnetic and structural 'nanoscale phase separation'[35]. Two closeby pemtons form a small stripe segment, while random charges generate intrinsic inhomogeneities: multipemton states of multiple scales occur, with spatially varying wandering ribbon-like strain/spin regions, or stripe-like bubbles from parallel opposite-sign ribbons depending on parameters. Because of the composite charge/spin/strain nature of the multiscale pemtons, there are both direct and cross responses of strain and magnetization to stress and magnetic field.

These intrinsic inhomogeneities are lattice deformations in response to random charges with isotropic profiles, but are far from random, or isotropic, themselves. They are not quenched-defect disorder from local lattice destruction, but annealed-texture complexity from local lattice preservation. They reflect the encoded unit-cell symmetry of the elastic-compatibility Fourier kernels, as seen in log-plots in Fourier space. The model provides a natural mechanism for temperature-dependent percolation of magnetic variables; for cross-responses of strain and spin to magnetic field and stress; and qualitatively explains other properties of complex electronic oxides[38]. We now present the GL model.

2. Ginzburg-Landau model

2.1. *Strain Free Energy*

We work with symmetry-adapted strains that are irreducible representations of the discrete symmetry group of the lattice unit cell. For a square-to-rectangle transition (that is a 2D version of the 3D tetragonal-to-orthorhombic transition), the OP deviatoric strain is $\varepsilon = (1/\sqrt{2})(E_{xx} - E_{yy})$. The non-OP strains are the compressional $e_1 = (1/\sqrt{2})(E_{xx} + E_{yy})$, and shear $e_2 = (E_{xy} + E_{yx})/2$ strains, respectively. The deviatoric OP strain $\varepsilon(\vec{r})$ represents rectangular deformations of a square unit cell in the x or y directions ($\varepsilon > 0$ or $\varepsilon < 0$), while the non-OP compression (expansion) corresponds to e_1 negative (positive). The invariant model free energy is

$$F = F^{(strain)}(\varepsilon, e_1, e_2) + F^{(mag)}(m) + F^{(coupling)}(n, m, \varepsilon, e_1), \quad (3)$$

and is scaled[24] so that all variables and parameters are dimensionless. The strain field contribution $F^{(strain)}$ is[22]:

$$F^{(strain)} = (a_0/2) \sum_r (\vec{\Delta}\varepsilon)^2 + F_{Landau}(\varepsilon) + F_{cs}(e_1, e_2). \quad (4)$$

Here $\vec{r}$ is a lattice site, $\vec{\Delta}$ is a discrete lattice derivative, and $\sqrt{a_0}$ is a strain variation length scale. The Landau term $F_{Landau} = (\tau - 1)\varepsilon^2 + \varepsilon^2(\varepsilon^2 - 1)^2$ has a triple-well in the deviatoric OP strain, that is moved by the scaled temperature $\tau = (T - T_{sc})/(T_s - T_{sc})$ to equal depth at a first-order transition temperature $T = T_s$, or $\tau = 1$. Below T_s , a Landau description of the transition gives an OP jump from $\varepsilon = 0$ to $\varepsilon = \bar{\varepsilon}(T)$. In the range $4/3 > \tau > 0$, the triple wells persist above and below the transition becoming double wells for $\tau < 0$ or $T < T_{sc}$. The non-OP compression/shear (cs) contributions are simply harmonic,

$$F_{cs} = \sum_r \frac{1}{2} A_1 e_1^2 + \frac{1}{2} A_2 e_2^2, \tag{5}$$

and would yield zero equilibrium values for non-OP strains, if these were independent scalar fields. Of course, Eq. (1) says they are not: effectively, $e_{1,2} = e_{1,2}(\varepsilon(\vec{r}))$. The harmonic non-OP free energy becomes $F_{cs}(e_1, e_2) \rightarrow F_{cs}(\varepsilon)$, and is the crucial texture-inducing long-range potential.

2.2. *Elastic Compatibility Potentials*

In 2D, the St. Venant constraint of Eq. (1) becomes[21,22,23,24]

$$C(\vec{r}) \equiv \vec{\Delta}^2 e_1 - \sqrt{8}\Delta_x \Delta_y e_2 - (\Delta_x^2 - \Delta_y^2)\varepsilon = 0, \tag{6a}$$

or in Fourier space,

$$C(\vec{k}) \equiv -\vec{k}^2 e_1(\vec{k}) + \sqrt{8} k_x k_y e_2(\vec{k}) + (k_x^2 - k_y^2)\varepsilon(\vec{k}) = 0. \tag{6b}$$

Thus for $k_x = \pm k_y$ diagonal OP twin boundaries, the non-OP strains $e_{1,2}$ can relax to zero in a Meissner-like expulsion. Conversely, if $e_{1,2}$ is forced to be nonzero, then nondiagonal OP textures must appear. Minimizing the free energy F_{cs} of Eq. (5) with respect to $e_{1,2}$ while maintaining the compatibility constraint,

$$\delta[F - \sum_{\vec{r}} \lambda(\vec{r}) C(\vec{r})] = 0 = \delta[F - \sum_{\vec{k}} \lambda(\vec{k}) C(\vec{k})], \tag{7a}$$

implying that

$$[A_1 e_1 + \lambda k^2]\delta e_1(\vec{k}) + [A_2 e_2 - \lambda\sqrt{8} k_x k_y]\delta e_2(\vec{k}) = 0, \tag{7b}$$

so the non-OP strains in Fourier space are obtained in terms of the Lagrange multiplier $\lambda(\vec{k})$. Inserting these into Eq. (6b) fixes the particular $\lambda(\vec{k}) \sim \varepsilon(\vec{k})$, that ensures compatibility is maintained. Hence the non-OP strains are expressed in terms of the OP strain, $e_{1,2}(\vec{k}) = B_{1,2}(\vec{k})\varepsilon(\vec{k})$, where $B_1(\vec{k}) = [(k_x{}^2 - k_y{}^2)k^2]/[k^4 + (8A_1/A_2)(k_xk_y)^2]$ and $B_2(\vec{k}) = -(A_1/A_2)\sqrt{8}k_xk_y/k^2)B_1(\vec{k})$. Substituting back into Eq. (5) yields[21,22]

$$F_{cs} = (A_1/2)\sum_k U(\hat{k})|\varepsilon(\vec{k})|^2; \quad U(\hat{k}) = [(k_x{}^2 - k_y{}^2)^2/[k^4 + (8A_1/A_2)(k_xk_y)^2]. \tag{8}$$

Clearly $U(\hat{k})$ depends on direction $\hat{k}$ but not magnitude $|\vec{k}|$ for long wavelengths (see below), and favours $\hat{k_x} = \hat{k_y}$ strain textures (such as diagonal domain walls). With $\hat{r}.\hat{r'} = \cos(\theta - \theta')$, in coordinate space, $F_{cs} = (A_1/2)\sum_{r,r'}\varepsilon(\vec{r})U(\vec{r} - \vec{r'})\varepsilon(\vec{r'})$, where the bulk compatibility potential $U(\vec{r} - \vec{r'}) \sim \cos 4(\theta - \theta')/|\vec{r} - \vec{r'}|^2$ falls off with a dimensional power just from phase space volume, and has the fourfold symmetry of the square unit cell. (The power law of course comes from compatibility, rather than proximity to some critical point.) We ignore for simplicity, both the finite-system surface[22] compatibility potential, $F_{cs}^{(surface)} \sim \sum_{k_x,k_y}|\varepsilon(k_x,k_y)|^2/|k_y|$, that yields 'true' or *equal*-width twins[39], and a tweed-inducing composition-fluctuation term[22] F_{compos}.

For $A_{1,2} << 1$, compatibility terms of Eq. (8) are energetically unimportant in F, and one expects simple polarons as from Eq. (2), with deformation of a single symmetry-broken ground state. However, for $A_{1,2} >> 1$, with a charge acting like a local stress, the non-OP energy costs may be large enough so it is profitable to locally give up symmetry-breaking, and generate opposite-sign strain, in adaptive elastic screening. Compatibility acting on such a multiple-minimum lattice, can then produce an unusual sign-varying (i.e. textured) extended polaron, with coupled fields like $m(\vec{r})$ also sign-varying, and modulated by $U(\vec{r} - \vec{r'})$, with anisotropic falloff. This excitation differs from a more familiar magneto-elastic polaron[31], that deforms a single ground state, and we will term it a *p*olaronic *e*lasto-*m*agnetic *t*exture, or 'pemton'.

As an aside, we note that since strains are (continuum) variables defined on a lattice dual to the unit cells, their discrete lattice derivatives in Fourier space are, strictly speaking, $\Delta_\mu{}^2 \to 4\sin^2(k_\mu/2)$. Such wavelength dependences reflect the discreteness of the underlying lattice, and occur naturally for example in the lattice Green's function or inverse Laplacian in Fourier space, $G(\vec{k}) = 1/[4\sin^2(k_x/2) + 4\sin^2(k_y/2)]$, that for long wavelengths depends only on magnitude $G(\vec{k}) \sim 1/|\vec{k}|^2$, and is independent of direction: it has an 'isotropic continuum' limit. The bulk compatibility potential U of course, has this trivial anisotropy from small-scale discreteness. However, by contrast, at long wavelengths it depends only on direction $\hat{k}$, and is independent of magnitude $|\vec{k}|$: it has an 'anisotropic scalefree continuum' limit. To emphasize the textures that derive from this crucial property of asymptotic anisotropy, we use the long-wavelength form given in Eq. (7), in our simulations.

2.3. *Couplings between Charge, Magnetization and Strain*

The model free energy for direct or staggered magnetization $m(\vec{r})$ is of a conventional form

$$F^{(mag)}(m) = \sum_{\vec{r}} f[(\frac{a_m}{2})(\vec{\Delta} m(\vec{r}))^2 + (T - T_{cm})m^2(\vec{r}) + \frac{1}{2}m^4(\vec{r})] - hm. \quad (9)$$

Here the zero-doping magnetic transition temperature is $T_{cm} < T_s$, h is a scaled magnetic field, and f is the ratio between magnetic and elastic energies. The symmetry-allowed couplings between the charge, spin, and strain variables are taken as

$$F^{(coupling)}(n, m, \varepsilon, e_1) = \sum_r A_{n,\varepsilon} n\varepsilon^2 + A_{nm} nm^2 + [A_{n1} ne_1 + A_{m1} m^2 e_1 + p_1 e_1], \quad (10)$$

where $p_1 > 0$ is an external compressional stress. We ignore for simplicity other symmetry-allowed terms like $m^2\varepsilon^2$, $m^2 {e_2}^2, n{e_2}^2$, etc.[40]

The dimensionless parameters are scaled in terms of the physical values, and can in principle be determined. Scaling[24] $FePd$ martensite[21] values yields strains ε, e_1, e_2 as fractions of a typical value 0.02; while the scaled stress $p_1 = 1$ (magnetic field $h = 1$) corresponds to $\sim$ 0.02 GPa ($\sim$ 1 Tesla); and $A_1 = 150, A_2 = 300$. We choose magnitudes and signs by physical arguments. The $A_{n\varepsilon} n\varepsilon^2$ term is like a local temperature $\sim \tau\varepsilon^2$, and for $A_{n\varepsilon} > 0$, says that the charge locally relaxes the strained unit cell towards its high-temperature $\varepsilon = 0$ symmetry[30]. The $A_{m1} m^2 e_1$ term acts like a local stress ($\sim m^2$) for e_1, and like a local temperature ($\sim e_1$) for m^2, and with $A_{m1} > 0$, magnetization is enhanced by a compressed ($e_1 < 0$) unit cell bringing spins closer. We choose illustrative model parameters as $A_1 = 50, A_2 = 105, \sqrt{a_0} = 0.5, \sqrt{a_m} = 1, \kappa = 2, A_{m1} = 5, T_s = 1, T_{sc} = 0.8, f = 0.3$.

The coefficient of the m^2 term in F, defines an effective local[35] transition temperature $T_{cm}^{eff}(\vec{r}) = T_{cm} - (A_{nm} n + A_{m1} e_1)/f$, and temperature deviation $\tau_{cm}(\vec{r}, T) \equiv f[T - T_{cm}^{eff}(\vec{r})]/A_{m1} = e_1 + [f(T - T_{cm}) + A_{nm} n]/A_{m1} \equiv \delta e_1(\vec{r}, T)$ that also defines a local shifted- strain deviation. Regions are below transition where they are 'sufficiently compressed',

$\tau_{cm}(\vec{r},T) = \delta e_1(\vec{r},T) < 0$, and as T rises, we expect the $\tau(\vec{r},T) > 0$ regions above transition with $m(\vec{r}) = 0$, to expand. This provides a mechanism for temperature-dependent magnetic percolation.

Obviously we cannot deal with the full quantum problem. Instead, we investigate texturing possibilities within a GL approach that follow from our generic model assumptions, and *mimic* cuprates and manganites by further model choices of parameter signs and interpretations of variables. The results could serve as a guide to microscopic calculations.

(i) For hole-doped 'cuprates' $n(\vec{r})$ is the local *hole* number density and $x_h =< n >$ the hole doping fraction. Summing over unit-normalized single-charge profiles at sites i, $n(\vec{r}) = \sum_i (\kappa^2/2\pi)e^{-\kappa|\vec{r}-\vec{r}_i|}$, with inverse decay length κ representing intersite tunneling. Here $m(\vec{r})$ is the *staggered* magnetization or 'chirality', and $A_{nm}nm^2$ for $A_{nm} > 0$ says the zero-doping Néel temperature $T_{cm} \neq 0$ decreases with x_h. The structural transition temperature also falls rapidly with doping[14] so we take $A_{n\varepsilon} >> 1$. Since holes repel lattice ions and expand unit cells, the $A_{n1}ne_1$ term has $A_{n1} < 0$, favouring $e_1 > 0$: $n(\vec{r})$ acts like a negative (expansive) local stress. The specific 'cuprate' parameter set is chosen as $T_{cm} = 0.6, A_{nm} = +1, A_{n\varepsilon} = 20, A_{n1} = -5$, and although illustrative, incorporates physical considerations above.

(ii) For 'manganites', we consider for conceptual simplicity a parent compound that is electron-doped through fraction $x_e =< n >$, with $n(\vec{r})$ now the electron number density. Here $m(\vec{r})$ is the *direct* (ferro-) magnetization density of the core spins, and the $A_{nm}nm^2$ term with $A_{nm} < 0$ says carrier electrons lock onto and align these core spins, mimicking the effects in microscopic models of a double exchange term, plus Hund's rule. We set the zero-doping value $T_{cm} = 0$. Thus an effective local Curie temperature $T_{cm}^{eff}(\vec{r}) = (-A_{nm}n - A_{m1}e_1)/f > 0$ emerges only through carriers and the compression they produce. (This provides a specific realization of the earlier speculations that manganites somehow have a local transition temperature[35].) Since metallicity shrinks unit-cell volume, we assume doped electrons effectively attract lattice ions, and take $A_{n1}ne_1$ term with $A_{n1} > 0$, favouring $e_1 < 0$: $n(\vec{r})$ is like a positive (compressive) local stress. The 'manganite' parameter set is $T_{cm} = 0, A_{nm} = -1, A_{n\varepsilon} = +2, A_{n1} = +5$.

It is worth relating our own approach to some previous approaches[21,27,33]. The classical work of St. Venant in 1864 is of course, in the textbooks[27], and was invoked by Baus and Lovett when considering a problem in defining surface tension, in terms of the fluid stress tensor. In an interesting paper[21], Kartha et al. performed an elastic displacement simulation, with a free energy sixth order in deviatoric displacement gradients (OP strain) and harmonic in the non-OP diplacement gradients, finding diagonal domain walls and tweed textures. In this $\vec{u}$-simulation, every term is anisotropic, and there is no explicit long-range term. In an effort to understand the observed ordering, they used the Baus-Lovett approach of treating strain as the variational variable, to minimize the non-OP harmonic free energy, subject to the St. Venant compatibility constraint, finding implicit long-range forces. Regarding our own work, we first performed OP strain relaxational simulations[22], that explicitly used the anisotropic long-range constraint forces in the square-to-rectangle free energy, with ε^6 now regarded as zeroth order in gradients, i.e. effectively a scalar, and the OP-OP potential now carrying all the anisotropy of the model. Every discrete symmetry transition has its own characteristic compatibility kernel, and the (static) kernels for all discrete 2D symmetries have been obtained[23]. A dynamic treatment for both over-and under-damped cases[24] shows that for some symmetries, the more general kernel $U(\hat{k}, \omega/|\vec{k}|)$ is dependent on both wavevector magnitude and frequency.

Turning to complex oxides in particular, the notion that lattice coupling is relevant, is of course, not new. Millis[2] suggested manganites might not be explainable in terms of double exchange alone, and that polarons could play a role, an idea taken up by many others, including Roeder et al.[31], who microscopically modelled magneto-elastic polarons, with a harmonic lattice. In the context of stripes in cuprates, Kusmartsev[32] has considered linear coupling of the displacement to a harmonic lattice plus isotropic Coulomb repulsion, and found multipolaronic electron strings can be stable. More recently, Khomskii and Kugel[33] have considered charges as in manganites as spherical or elliptical 'Eshelby inclusion' defects in a harmonic elastic medium, and invoked defect-defect anisotropic long-range interaction. The defect forces are regarded as unscreened (except for surface images); the quadrupolar modulations are of a single ground state structure; and although strain tensor components are implicitly present in the analysis, compatibility is not explicitly considered. An effective next-nearest-neighbour anisotropic Ising model is proposed, with discussions of possible stabilization of complex structures in manganites. Our own approach considers charges as local stresses or temperatures that do not cause defects, and explicitly couple locally (in symmetry-allowed ways) to the strain tensor components. There is an OP-OP anisotropic long-range potential, but no charge-charge interaction postulated at the outset: this effective interaction emerges from the coupled dynamics. The integrity of the lattice through St. Venant compatibility is the central focus, throughout. The key idea of adaptive elastic screening is introduced, and quadrupolar modulations are sign-variations of the OP strain, denoting the presence of two degenerate structures. This is seen in simulations, to which we now turn.

3. Simulation of Textures

Since non-OP strains $e_{1,2}$ are compatibly related to ε, the free energy is $F = F(n, m, \varepsilon)$, with Eq. (9) becoming *nonlocal* in the couplings of n, m, p_1 to ε. We find the equilibrium textures through a relaxational equation, (that is the overdamped limit of a general ferro-elastic dynamics[24]):

$$\dot{\varepsilon} = -\frac{\partial F}{\partial \varepsilon}; \quad \dot{m} = -\frac{\partial F}{\partial m}; \tag{11}$$

We consider NxN systems with $N = 128$ and periodic boundary conditions (using, therefore, charge profiles defined through their site Fourier transform). Our parameters corresponds to a regime of weak magnetism, strong electron-phonon coupling, and dominant compatibility forces. A 'parent compound' is first prepared with $\varepsilon(\vec{r})$ and then $m(\vec{r})$, relaxed from initial conditions random in space around meanfield values, with up to $\sim 10^4$ time steps of size $\Delta t \sim 0.01$. We then add the charges of concentration $x =< n >$, at fixed positions, randomly (but forbidding double occupancy). Runs are up to 3×10^4 time steps, using the temporal energy-flattening, magnetic and strain forces, averages and max/min values of variables, as diagnostics for equilibration[24]. All plots subtract the parent-compound background. Both $\vec{k}$ and $\vec{r}$ space plots are needed for a full understanding, see e.g., Ref. [36].

Figure 1 gives plots for 'cuprate' parameter set and a single charge. Here the Néel temperature is $T_{cm} = 0.6$ and so the parent compound has a uniform staggered magnetization. With this background subtracted, contour plots are shown of the deviatoric $\varepsilon(\vec{r})$, and compressional $e_1(\vec{r})$ strain and staggered magnetization $m(\vec{r})$; and Fourier space plots of $|\varepsilon(\vec{k})|^2$, $|e_1(\vec{k})|^2$, and $|m(\vec{k})|^2$. Note the lobed, quadrupole-like texturing in coordinate space, extending in over a score of lattice spacings, much larger than the unit-cell extent of one of the charges $n(\vec{r})$ seen in Fig.1. There are both signs of chirality and strain (with spatial averages close to zero). This concretely illustrates the concept[35] of 'nano-scale phase separation'. Since the ne_1 local stress term would by itself produce a bare single-sign strain, clearly

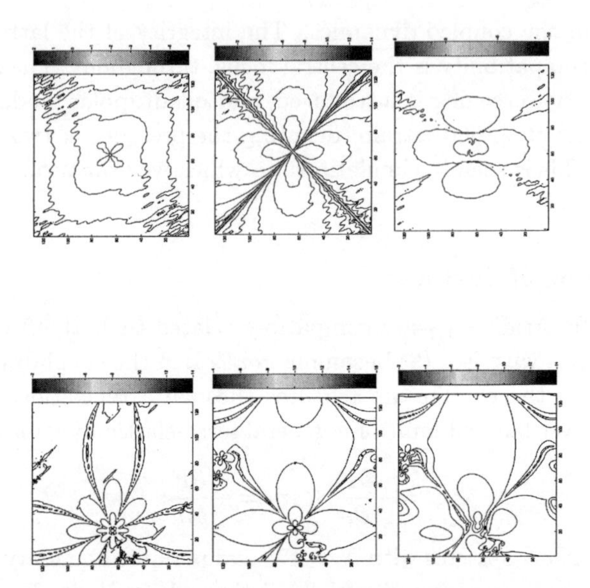

Figure 1. *Single polaronic elasto-magnetic texture or pemton state for 'cuprate' case:* Scaled temperature $T = 0.5$. Background parent compound, with OP twins and Néel transition temperature $T_{cm} = 0.6$, is subtracted. Panels read left to right. Bottom panel: Contour plots of deviatoric $\varepsilon(\vec{r})$, compressional $e_1(\vec{r})$ strains, and staggered magnetization $m(\vec{r})$. Top panel: Log-scale Fourier plots of $|\varepsilon(\vec{k})|^2$, $|e_1(\vec{k})|^2$, and $|m(\vec{k})|^2$. Grey scale number bars give an idea of magnitudes.

adaptive elastic screening is occuring to mix in both ground states of the nonlinear lattice into a strain- and spin-neutral solitonic texture.

Figure 2 shows the 'cuprate' multipemton strain and staggered magnetization, for random doping $x_h = 0.1$, showing that parallel opposite-sign strain ribbons form bubbles of stripes, in the deviatoric strain $\varepsilon(\vec{r})$ or '$2D$ orthorhombicity' (that also show up in $m(\vec{r})$). The neighbouring column of log-scale Fourier plots shows the signature of the compatibility kernel of Eq. (8): there is a four-lobe basic structure, as in the single-pemton case, with many crinkles, that on close inspection have an inversion ($\vec{k} \to -\vec{k}$) symmetry connecting most of even the fine details. The randomness of the doping has modified, but not destroyed, the hidden order of a compatible lattice. We do not show charges in an array[4], but note that from momentum conservation[37] on dominant wavevectors in the couplings

$\sim \sum_r nm^2, \sum_r ne_1, \sum_r m^2 e_1$, the staggered magnetization m has wavevector $2k_m = k_n = k_1$, half that of the charge lines n, that would be anti-phase boundaries for the chirality.

Figure 2. *Multipemtons for 'cuprate' parameters:* Scaled temperature $T = 0.5$, random hole concentration $x_h = 0.1$. The parent compound background is subtracted. Panels read left to right. Bottom panel: Contour plots of deviatoric $\varepsilon(\vec{r})$, compressional $e_1(\vec{r})$ strains, and staggered magnetization $m(\vec{r})$. Top panel: Log-scale Fourier plots of $|\varepsilon(\vec{k})|^2$, $|e_1(\vec{k})|^2$, and $|m(\vec{k})|^2$, showing the multiscale nature of texturing.

For 'manganites', as in Fig. 3, the pemtons are broader and floppier, but retain the generic 'butterfly' shape, in strain and ferromagnetic moment $m(\vec{r})$. (Note here the Curie temperature $T_{cm} = 0$ in the parent compound, and all local magnetization appears only due to the local internal stresses of the charges and strains.) Two pemtons deform each other, and when closeby, form small stripe segments of alternating strain signs. For increased doping at fixed temperature, the average magnetization rises sharply above $x_e = 0.13$ to $< m >= -0.21$ at $x_e = 0.15$. These apparently random ribbon-like multipemton states are shown in Fig. 4. The Fourier plots again reveal the hidden order, and fine-scale inversion symmetry of most details. The plots are reminiscent of diffuse X-ray and neutron scattering[12]; the diagonal streak in $|\varepsilon(\vec{k})|^2$ is due to the charges affecting the diagonal quasi-twins in the parent background.

Application of external long wavelength magnetic fields and stresses change both strains and magnetizations, with the cross-responses and cloud-like gradations[9] demonstrating the *composite* charge/spin/strain nature of the pemton states as presented elsewhere[38]. The scenario that emerges is of inhomogeneous ribbon-like percolating regions, with small average values of strain and magnetization emerging from a fine balance between large local opposite-sign values, that is tipped by applied uniform stress or magnetic field. This suggests a qualitative understanding of CSR and CMR effects, in terms of the "compressed $\sim$ magnetic $\sim$ metallic" connections. The field-dependent high/low conductivity regions seen in STM/TEM pictures[9,10,11] would then correspond to channels of compressed/expanded unit-cell regions (increased/decreased tunneling), or oriented/disordered spin regions (conduction favoured/disfavoured through double-exchange).

4. Conclusion

In summary, these conceptually illustrative Ginzburg-Landau simulations show that charges doped into magnetic background in a nonlinear elastic lattice with displacive first-order structural transitions, can produce polaronic elasto-magnetic textures or 'pemtons', with extended sign-varying spin and strain regions, i.e. nanoscale phase separation. The model provides a qualitative understanding of 'stripe' formation in cuprates; and of manganite 'colossal' effects with direct and cross-coupling of strain and magnetization to stress and magnetic field; with temperature-dependent magnetic percolation. The reason why the proposed excitations could be relevant for complex electronic oxides, is that atomic bases of perovskite octahedra (with directionally bonded transition-metal ions, and polarizable/deformable corner oxygens[1,2]) can support large unit-cell deformations through tilts and bond angle bucklings, without bond breaking.

Of course this simple $2D$ GL model with immobile, fixed-profile charge densities, cannot be expected to reproduce the complex phase diagram of manganites. The rich variety of charge- (or pemton-) ordered states clearly could arise from the orienting compatibility forces, and the model could be extended so charges are allowed to hop. If charge profiles are

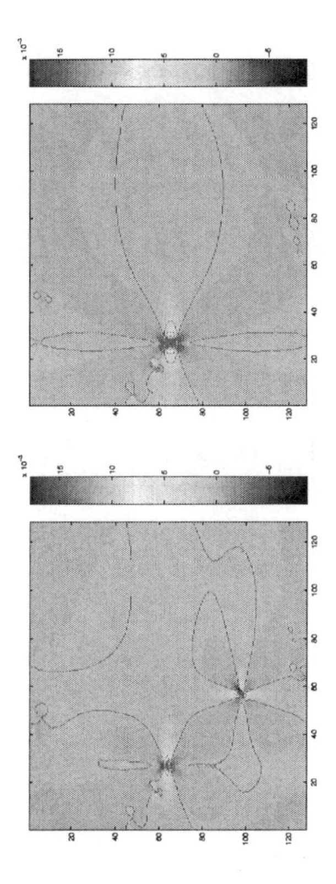

Figure 3. *'Manganite' bipemton states:* Grey scale: $e_1(\vec{r})$ for separated (bottom) and closeby (top) pemtons. Note formation of strain/spin stripe segments.

allowed to relax into (compressed) strain lobes, then orbital ordering is another natural consequence. Simulations could be done of the actually[25] 3D structures and Jahn-Teller distortions. Microscopic modelling could include variables related to strain tensors e.g. plane bucklings[41] induced by octahedral tilts. Further experimental work could include STM studies on manganites/cuprates, processing the data in terms of strain-like combinations, to uncover signatures of pemtons, and their responses to external control fields.

It is instructive to think of levels of description of spatial variation. Level 0 is the mechanics scale of system size and system shape, e.g. for macroscopic shape memory effects. Level 1 ranges from system size to unit-cell scales, and is the arena of strain variations and multiscale texturing, that here are constrained by compatibility. Level 2 is the intracell

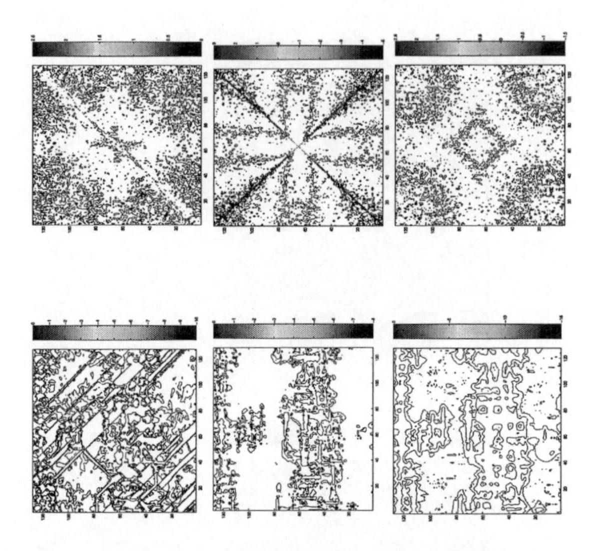

Figure 4. *'Manganite' multipemton states:* Scaled temperature $T = 0.5$ and electron doping fraction $x_e = 0.1$. Panels read left to right. Bottom panel: Contour plots of deviatoric $\varepsilon(\vec{r})$, compressional $e_1(\vec{r})$ strains, and magnetization $m(\vec{r})$. Top panel: Log-scale Fourier plots of $|\varepsilon(\vec{k})|^2$, $|e_1(\vec{k})|^2$, and $|m(\vec{k})|^2$, showing the multiscale nature of texturing.

world[42] of 'microstrain' deformation of atomic bases, e.g. octahedral tilts. Level 3 is the Angstrom-scale regime of electronic structure, where implicit adiabaticity assumptions of conventional approaches may have to be relaxed, to fully capture the mutual accommodations of dopant charges and ionic optimizations[43]. The levels are linked, of course, if multiscale lattice integrity holds. In particular, the Level 1/Level 2 interface is where the elastic physics meets the structural chemistry. The key insight from our Level 1 model is that intrinsic inhomogeneities are not random, quenched-defect, uncorrelated disorder. Rather, they are multiscale annealed-texture responses to doping perturbation, of an unruptured, nonlinear lattice; with a hidden, discrete-symmetry order from their structural compatibility potential.

We thank Seamus Davis, Carlos Frontera, and Venkat Pai for useful conversations. This work was partly supported by the U.S. Dept. of Energy.

References

1. *Stripes 2000*, eds. N.L. Saini and A. Bianconi, Int. J. Mod. Phys. **14**, 3289 (2000).
2. *Colossal Magnetoresistance and Related Properties*, eds. B. Raveau and C.N.R. Rao, (World Scientific, Singapore, 1998); A.J. Millis, Nature **392**, 147 (1998).
3. Z. Nishiyama, *Martensitic Transformations*, (Academic, New York, 1978); L.E. Tanner, J.Phys. (Paris), Colloq, **43**, C4-169 (1982); E.K.H. Salje, *Phase Transitions in Ferroelastic and Coelastic Solids*, (Cambridge University Press, Cambridge, UK, 1990).
4. J.M. Tranquada, B.J. Sternleich, J.D. Axe, Y. Nakamura, and S. Uchida, Nature **375**, 561 (1995); X.J. Zhou, P. Bogdanov, S.A. Kellar, T. Noda, H. Eisaki, S. Uchida, Z. Hussain, and Z.-X. Shen, Science **286**, 268 (1999); H.A. Mook, P. Dai, and F. Dogan, Phys. Rev. Lett. **88**, 097004 (2002).
5. A. Bianconi, N.L. Saini, A. Lanzara, M. Missori, T. Rosetti, H. Oyanagi, H. Yamaguchi, K. Oka, and T.Ito, Phys. Rev. Lett. **76**, 3412 (1996); A. Bianconi, N.L. Saini, S. Agrestini, D. DiCastro, and G. Bianconi, Int. J. Mod. Phys. B **14**,3342 (2000).
6. J.-X. Liu, J.-C. Wan, A.M. Goldman, Y.C. Chang, and P.Z Jiang Phys. Rev. Lett. **67**, 2195 (1991).
7. J.E. Hoffman, K.McElroy, D.-H. Lee, K.M. Lang, H. Eisaki, S. Uchida, and J.C. Davis, Science **297**, 1148 (2002); D.J. Derro, E.W. Hudson, K.M. Lang, S.H. Pan, J.C. Davis, J.T. Markert, and A.L. deLozanne, Phys. Rev. Lett. **88**, 097002 (2002).
8. Y. Hwang, T.M. Palstra, S.-W. Cheong, and B. Batlogg, Phys. Rev. B **52**, 15046 (1995).
9. M. Faeth, S. Friesen, A.A. Menovsky, Y. Tomioka, J. Aaarts, J.A. Mydosh, Science **285**, 1540 (1999).
10. V. Podzorov, C.H. Chen, M.E. Gershenson and S.-W. Cheong, Europhys. Lett. **55**, 411 (2001).
11. M. Uehara, S. Mori, C.H. Chen, and S.-W. Cheong, Nature **399**, 560 (1999); M. Uehara and S.-W. Cheong, Europhys. Lett., **52**, 674 (2000).
12. L. Vasiliu-Doloc et al., J.W. Lynn, J. Mesot, O.H. Seeck, J.F. Mitchell, Phys. Rev. Lett. **83**, 4393 (1999); S. Shimomura, N. Wakabayashi, H. Kuwahara, and Y. Tokura, Phys. Rev. Lett. **83**, 4389 (1999).
13. A.N. Lavrov, S. Komiya, and Y. Ando, Nature **418**, 385 (2002); S.B. Ogale et al., Phys. Rev. Lett. **77**, 1159 (1996).
14. A.S. Alexandrov and N.F. Mott in *Polarons and Bipolarons in High-T_c Superconductors*, eds. E.K.H. Salje, A.S. Alexandrov, and W.Y. Liang, (Cambridge University Press, Cambridge, UK, 1995).
15. V.K. Wadhawan, Phys. Rev. B **38**, 8936 (1988); Ferroelectrics **97**, 171 (1989).
16. J.A. Krumhansl in *Lattice Effects in High-T_c Superconductors*, eds. Y. Bar-yam, T. Egami, J. Mustre-de Leon, and A.R. Bishop, (World Scientific, Singapore, 1992); S. Sergeenko and M. Ausloos, Phys. Rev. B **52**, 3614 (1995); L. Yiping, A. Murthy, G.C. Hadjipayanis, and H.W. Wan, Phys. Rev. B **54**,

3033 (1996).
17. O. Tikhomirov, H. Jiang, and J. Levy, Phys. Rev. Lett. **89**, 147601 (2002).
18. G.-M. Zhao, K.K. Singh, and D.E. Morris, Phys. Rev. B **50**, 4112 (1994); D. Zech et al., K. Conder, E. Kaldis, E. Liarokapis, and Nature **371**, 681 (1994); D. Rubio Temprano, J. Mesot, S. Janssen, K. Conder, A. Furrer, H. Mutka, and K.A. Müller, Phys. Rev. Lett. **84**, 1990 (2000).
19. G.-M. Zhao and K.A. Müller, Nature **381**, 676 (1996).
20. A.E. Jacobs, Phys. Rev. B **31**, 5984 (1985); **61**, 6587 (2000).
21. S. Kartha, J.A. Krumhansl, J.P. Sethna, and L.K. Wickham, Phys. Rev. B **52**, 803 (1995).
22. S.R. Shenoy, T. Lookman, A. Saxena, and A.R.Bishop, Phys. Rev. B, **60**, R12537 (1999).
23. D.M. Hatch, T. Lookman, A. Saxena, and S.R. Shenoy, Phys. Rev. B. **67**.
24. T. Lookman, S.R. Shenoy, K.Ø. Rasmussen, A. Saxena, and A.R. Bishop, Phys. Rev. B **67**, 024114 (2003).
25. K.Ø. Rasmussen, T. Lookman, A. Saxena, A.R. Bishop, R.C. Albers, and S.R. Shenoy, Phys. Rev. Lett. **87**, 055704 (2001).
26. M. Seul and D. Andelman, Science **267**, 476 (1995); K.-O. Ng and D. Vanderbilt, Phys. Rev. B **52**, 2177 (1995).
27. M. Baus and R. Lovett, Phys. Rev. Lett. **65**, 1781 (1990); *ibid* **67**, 407 (1991); Phys. Rev. A **44**, 1211 (1991); D.S. Chandrasekharaiah and L. Debnath, p. 218, *Continuum Mechanics*, (Academic, San Diego 1994).
28. J. Zaanen and O. Gunnarson, Phys. Rev. B **40**, 7391 (1989).
29. S.A. Kivelson and V.J. Emery, Physica C **235**, 189 (1999).
30. A.J. Millis, Phys. Rev. B **53**, 8434 (1996).
31. H. Roeder, J. Zhang, and A.R. Bishop, Phys. Rev. Lett. **76**, 1356 (1996).
32. F. Kusmartsev, Phys. Rev. Lett. **84**, 5026 (2000); Europhys. Lett. **57**, 557 (2002).
33. D.I. Khomskii and K.I. Kugel, Europhys. Lett. **55**, 208 (2001).
34. M.L. Kulic, Phys. Rep. **338**, 2 (2000).
35. J. Burgy et al., Phys. Rev. Lett. **87**, 277202 (2001); M.B. Salamon, P. Lin and S.H. Chun, Phys. Rev. Lett. **88**, 197203 (2002); E. Dagotto, T. Hotta, and A. Moreo, Phys. Rep. **344**, 1 (2000)
36. A.L. deLozanne, this volume.
37. O. Zachar, V.J. Emery and S.A. Kivelson, Phys. Rev. B **57**, 1422 (1998).
38. A.R. Bishop, T. Lookman, A. Saxena, and S.R. Shenoy, submitted.
39. G.R. Barsch, B. Horovitz, and J.A. Krumhansl, Phys. Rev. Lett. **59**, 1251 (1987).
40. A. Asamitsu, Y. Moritomo, R. Kumai, Y. Tomioka, and Y. Tokura, Phys. Rev. B **54**, 1716 (1996).
41. S.R. Shenoy, V. Subrahmanyam, and A.R. Bishop, Phys. Rev. Lett. **79**, 4657 (1997); D. Mihailovic, V. V. Kabanov, and K. A. Müller, Europhys. Lett. **57**, 254 (2002).
42. C. Zener, Phys. Rev. **71**, 846 (1947).
43. K. Iwano, Phys. Rev. B **64**, 184303 (2001); J.A.M. McAllister and J.P. Attfield, Phys. Rev. B **66**, 04514 (2002).

SUPERCONDUCTIVITY AND THE STRIPE STATE OF TRANSITION METAL OXIDES

A. H. CASTRO NETO
Department of Physics, Boston University
590 Commonwealth Ave., Boston, MA, 02215

The interplay between charge order and superconductivity has been a subject of intense debate in the field of high temperature superconductors. While the experimental data seems to indicate that charge order competes with superconductivity, it is accepted that in the underdoped regime of the cuprates, where the superconducting state is born out of an antiferromagnetic Mott insulator state, there is a strong tendency to the formation of inhomogeneous stripe states. We are going to discuss a possible scenario for the appearance of such a superconducting state as an interplay between magnetic and vibrational degrees of freedom. We propose that due to the multiscale structure intrinsic to these systems, stripes and d-wave hole bound states co-exist in the presence of phonons. We also show that as a result of this co-existence a superconducting state with multiple topological excitations is created. We discuss the nature of this superconducting state and its experimental consequences.

1. Introduction

One of the main problems in condensed matter theory since the discovery of high temperature superconductors in 1986 [1] is related with the possible dilute phases of Mott insulators [2]. At half-filling high temperature superconductors are insulating two-dimensional (2D) antiferromagnets well described by the isotropic Heisenberg model [3] with a large charge transfer gap. These are trademarks of mottness [4]. These characteristics lead Anderson [5] to propose that cuprates can be well described by a Hubbard model with very large inter-site repulsion U. Later studies showed that close to half-filling and infinite U the model maps into the $t-J$ model:

$$H = -t \sum_{<i,j>,\alpha} P c^{\dagger}_{i,\alpha} c_{j,\alpha} + J \sum_{<i,j>} \mathbf{S}_i \cdot \mathbf{S}_j \tag{1}$$

where t is the hopping energy and $J \approx t^2/U \ll t, U$ is the exchange interacting between neighboring electrons spins, $\mathbf{S}_i = c^{\dagger}_{i,\alpha} \vec{\sigma}_{\alpha,\beta} c_{i,\beta}$ ($c_{i,\alpha}$ is the electron annihilation operator at the site i with spin projection $\alpha = \uparrow, \downarrow$, and σ^a with $a = x, y, z$ is a Pauli matrix). In Eq.(1) P is the projection operator into single site occupancy (double occupancy is forbidden). A nice feature of (1) is that it reduces trivially to the Heisenberg model at half-filling.

Density matrix renormalization group (DMRG) [6] calculations of the t-J model on ladders have indicated that the ground state is striped, that is, the system is characterized by doped anti-phase domain walls between two different staggered configurations of the Neel state. This result is consistent with the experimental observation of stripe states in cuprates [7]. These calculations, however, are very sensitive to boundary conditions and also the inclusion of other terms in the Hamiltonian. The introduction of next nearest neighbor hoping t' in the problem can favor mobile d-wave pairs of holes for $t' > 0$ and single hole excitations (spin-polarons) for $t' < 0$ [8]. At $t' = 0$ the stripe state is very close in energy to the d-wave pair state, thus, a small change in the boundary conditions or inclusion of small perturbations in the Hamiltonian can easily favor one many-body state relative to another. The quasi-degeneracy between different many-body states is an important characteristic of strongly correlated systems and that is what makes the problem rather complicated. Moreover, in real systems other effects can be responsible for the selection of the many-body state, that is, to the lift of the quasi-degeneracy of the many-body states. Lattice anisotropies, for instance, can easily select the stripe state compared to the pair state [9,10].

Although the DMRG calculations were originally performed in a SU(2) t-J model the same physics is essentially true in the Ising t-J model (the so-called t-J_z model). The reason for the similarity between the SU(2) t-J (where the spins have dynamics) and the Ising t-J (where the spins are static) can be understood on the basis of the fluctuating time of each component in the problem. In the SU(2) model spins fluctuate with a rate $\tau_s \approx \hbar/J$ while the time scale for hole motion is $\tau_h \approx \hbar/t$. When $J/t < 1$ (the physical regime of the model) one has $\tau_s > \tau_h$, that is, the holes move "faster" than spins. In this case a Born-Oppenheimer approximation is reasonable and the holes move in the static configuration of the spins. Since the spins have slow dynamics the two problems become essentially identical [11]. The great advantage in working with the t-J_z model is that many of its properties can be easily studied numerically and calculated analytically.

Consider an infinite anti-phase domain wall along one of the crystalline directions of the system (see Fig.1). The cost in energy per hole to create such a state is $J/2(n^{-1} - 1)$ where n is the linear density of holes along the stripe. However, the hole acquires translational invariance along the stripe and therefore has a gain of energy of kinetic $-2t\sin(\pi n)/(\pi n)$ due to the longitudinal hopping. For $J/t = 0.4$ one can show that the energy is minimized for $n = 0.32$ with energy $E_b \approx -1.255t$ that is larger than the energy of the hole in the bulk (spin polaron), $E_{sp} = -2\sqrt{3}t \approx -2.37t$ [11]. The problem here is that we have not included the transverse motion of the holes perpendicular to the stripes. This motion further reduces the kinetic energy of the system but also gives a finite width to the hole wavefunction. Using a retraceable path approximation (but ignoring hole-hole interactions) one can calculate analytically the Green's function for the holes [14].

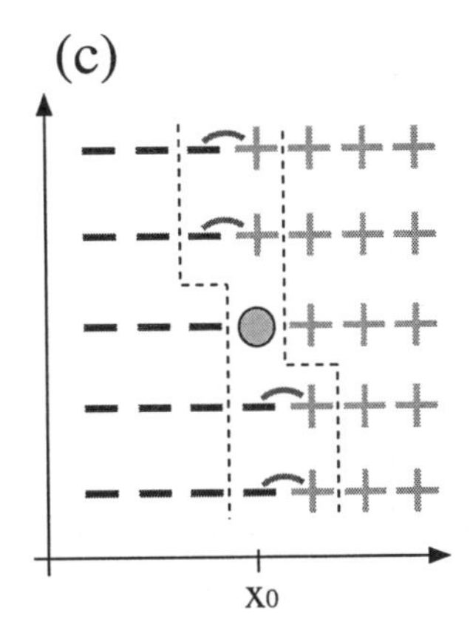

Figure 1. Anti-phase domain wall generated by one hole in the t-J_z model. The plus (minus) sign labels a classical staggered magnetization state.

The holes are confined to the anti-phase domain wall by the potential generated by strings of overturned spins (that is, there is a linearly growing potential transverse to the stripe direction). One can also show that in this configuration the hole is actually a holon, that is, it does not carry spin but charge and any motion of the hole away from the stripe produces a spinon with energy J. In the bulk the hole carries both spin and charge and therefore is a spin polaron. Thus, there is a local spin-charge separation although there is no macroscopic spin-charge separation [14]. Because of this effect Trugman loops [12], which are responsible for the hole confinement in the absence of anti-phase domain walls, are not effective since the motion of the holes away from the wall always produces an excitation of finite energy.

One finds that for $J/t = 0.4$ and $n \approx 0.3$ the energy of the stripe state is $E_0 \approx -2.5t$ and therefore lower than the energy of the spin polaron. Furthermore, the width of the hole wavefunction is of order of 3 to 4 lattice spacing [14] and therefore extends quite a large distance from the anti-phase domain wall in contrast with the usual cartoon picture where the stripe is only one lattice spacing wide [7]. Thus, to think of a stripe as a completely one-dimensional object is rather misleading since the hole makes long incursions into the antiferromagnet. Moreover, it is clear from these analytic calculations that it is the single hole kinetic energy which is responsible for the stabilization of the stripe state.

We should stress that we are discussing the ground state, that is, the lowest energy stationary state of the problem and therefore the concept of "stripe fluctuations" refers, in this context, to excitations that are separated from the ground state by an energy of order $t(J/t)^{2/3}$ because of confinement in the transverse direction [14]. These results are not only consistent with the DMRG results for the SU(2) t-J model for small J/t [13] but in complete agreement with DMRG results for the t-J_z model [15]. As a consequence the stripe is metallic in contrast with Hartree-Fock results that always produce a gap [16]. As in a Luttinger liquid [17] one expects that hole-hole interactions drive the system toward a CDW phase that can be easily made insulating by the presence of any amount of disorder [18]. Moreover, interactions make the density of states to vanish at the chemical potential with a non-universal power of the interaction parameter. Clearly in this picture the stripes do **not** superconduct, in fact they have a strong tendency to behave in an insulating way.

2. The model

Our starting point is the assumption that in the presence of electron-electron interactions alone the stripe array is the ground state of the problem. This state is essentially insulating and has no superconducting correlations (in accord with the Luttinger liquid physics and our calculations in the t-J_z model [14]). We model the stripes by a Luttinger liquid that is 1/4 filled and in a CDW state with zero density of states at the Fermi energy, μ (see Fig.2). We also assume (in accordance with the DMRG results for ladders) that a *bosonic* state of d-wave pairs exist below the chemical potential. This state is empty because it is a state of holes and the chemical potential is above it. The energy required for taking two electrons from the bosonic state and putting at the stripe Fermi energy is $2\mu - E_b$. Thus, with only electron-electron interactions the ground state has no pair and only insulating stripes.

In order to obtain any kind of non-trivial ground state the bound state of holes has to be excited to the stripe Fermi surface by a new particle. By conservation of energy this excitation requires an energy $2\mu - E_b$ and therefore one needs a massive, high energy, particle. Moreover, this particle

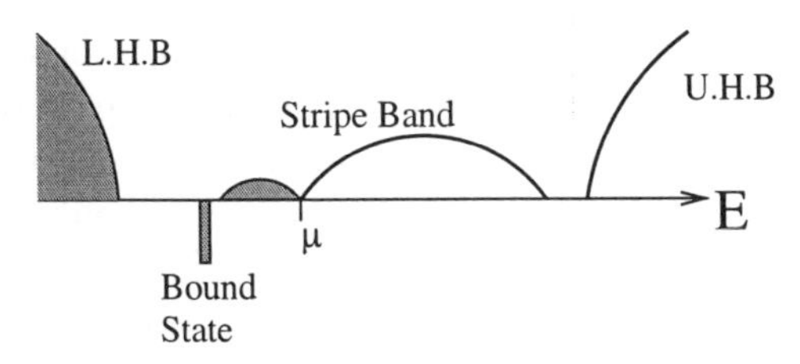

Figure 2. Density of states in the model: U.H.B. is the upper Hubbard band and L.H.B. is the lower Hubbard band.

has to be spinless since the d-wave pair is in a singlet state. Thus, we immediately eliminate para-magnons from the picture since these have spin one. The most easily accessible excitation in the system are optical phonons. Conservation of momentum, however, imposes a tough constraint. Along the stripe the fermions have their momentum in the direction parallel to the stripe (say the x-direction). In order for d-wave pairs with momentum (p_x, p_y) to decay into two fermions with opposite momentum $(+k_F, 0)$ and $(-k_F, 0)$ (where $k_F = \pi n/2$) and produce a particle with momentum (k_x, k_y) one has to require: $p_x = k_x = 0$ and $p_y = k_y$. One can show that because of locality of the interaction the coupling constant is a monotonic increasing function of p_y and therefore the larges coupling is at the Brillouin zone edge, that is, at $k_y = \pi/a$ [19]. This is also consistent with the fact that one needs an optical phonon at the edge of the zone. Notice that conservation of momentum and energy and the locality of the interaction requires that the largest coupling between stripes and d-wave pairs occurs in the direction **perpendicular** to the stripe direction and not along it. This picture has been also observed in numerical studies of spin-fermion models by Moreo and collaborators [20].

This scenario can be directly contrasted with other theories [21] that assume, from the very beginning, that the conduction occurs in the direction along the stripes and that the stripes, by the proximity to the antiferromagnet have a spin gap (pairing). In these theories the coupling between stripes occurs via the Josephson coupling of pairs by direct tunneling [22]. In our case, one has propagating d-wave pairs, that is, the problem is not one-dimensional like in ref.[21] but instead is two-dimensional but anisotropic. Furthermore, our scenario requires the lattice dynamics to be included from the very beginning otherwise, as explained above, the ground state would be insulating and therefore uninteresting. These two scenarios can be distinguished experimentally, as we explain above.

On the one hand, as a consequence of the finite momentum coupling at $\mathbf{Q}_L = (0, \pi/a)$ the lattice distorts to provide the momentum for the superconducting state to develop. On the other hand, superconductivity appears first at finite momentum in the direction perpendicular to the stripes. The appearance of superconductivity at $\mathbf{Q}_L$ can be associated with the opening of the pseudo-gap observed in angle resolved photo-emission experiments (ARPES) at these points in the Fermi surface [23]. Moreover, the phonon anomalies that are observed at the Brillouin zone edge in phonon density of states measurements [24] and more recently in ARPES [25] can be associated with the mechanism of pairing.

Another interesting consequence of the mechanism is related with the problem of phase coherence. The condensation of the d-wave pairs can be described in terms of the pair bosonic operator $B_{\mathbf{Q}_L}$ and the lattice distortion in terms $u_{\mathbf{Q}_L}$. Long range order requires that $\Delta = \langle B_{\mathbf{Q}_L} \rangle \neq 0$ and $u = \langle u_{\mathbf{Q}_L} \rangle \neq 0$. However, since these order parameters are complex we can always write:

$$
\begin{aligned}
\Delta &= |\Delta| e^{i\varphi_S} \\
u &= |u| e^{i\varphi_L}
\end{aligned} \tag{2}
$$

where $|\Delta|$ and $|u|$ are the amplitude of the order parameters and $\varphi_{S,L}$ their phases. Because the problem is essentially two-dimensional and the superfluid density is low one expects the phases to fluctuate strongly leading to a non-conventional phase coherence process. Thus, even if the amplitudes of the order parameters are finite one has to require that $\langle e^{i\varphi_{S,L}} \rangle \neq 0$. Moreover, long-range order in one of the order parameters does not necessarily imply long-range order in the other although it certainly require that the amplitudes are finite.

The simplest case to consider is the finite temperature where quantum fluctuations are irrelevant. In this case it is straightforward to show that the action of the problem can be written as [19]:

$$
S = \frac{\beta}{2} \int d^2x \left\{ \sum_{\alpha=S,L} \frac{\sigma_\alpha}{M_\alpha} \left(\nabla \varphi_\alpha(\mathbf{r})\right)^2 + \frac{v_F}{2\pi N a} \left(\sum_{\alpha=S,L} \partial_x \varphi_\alpha(\mathbf{r}) \right)^2 \right\} . \tag{3}
$$

where β is the inverse temperature, σ_α and M_α are the stiffness and the effective masses, respectively, v_F is the Fermi velocity of the stripe band, and Na is the distance between stripes (a is the lattice spacing). The consequence of this model is that topological defects of the superconductor (that is, superconducting vortices) are directly coupled to topological defects of the lattice (that is, dislocation loops). Thus, fluctuations in the lattice in the normal state can produce vortex-anti-vortex pairs and vice-versa. In fact, there are reports of the observation of vortices in the normal phase by Ong and collaborators [26]. It would be very interesting to try to relate experimentally the observation of vortices with the observation of phonon anomalies.

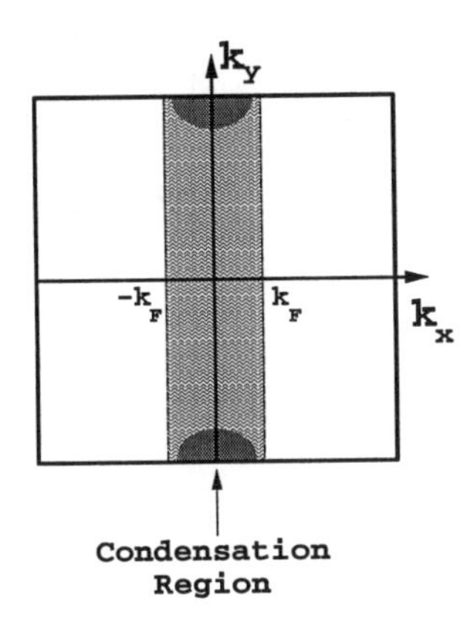

Figure 3. Mono-domain Brillouin zone.

3. π-Junctions and Stripe Domains: the Resonance Peak

All that has been said so far refers to a single stripe array with all the stripes oriented along the x-direction. In momentum space the situation is summarized in Fig.3. One sees that the d-wave pairs condense at $(0, \pm\pi/a)$. This condensation induces a superconducting gap at the stripe Fermi surface [19]. Notice that because of the flat nature of the Fermi surface the gap state is of the s-wave type. However, this s-wave state can be either positive or negative relative to the sign of the pair wavefunction. Thus, it is not possible to generate an d-wave state out of an array of infinitely long stripes. To get a true d-wave superconductor one has to rely on the presence of stripe domains.

If one consider two orthogonal domain of stripes one notes that it is possible to get four different configurations [27]. If the stripes have the same sign (either negative or positive) one obtains an anisotropic s-wave (see Fig.4(a)). If however, a π-junction is established between stripes [28] a d-wave superconductor can emerge as shown in Fig.4(b). Furthermore, with equal number of stripe domains the pseudo-gap and the lattice anomalies will occur with the same intensity at four points in the Brillouin zone, namely, $(\pm\pi/a, 0)$ and $(0, \pm\pi/a)$. In this case the scenario where the pseudo-gap occurs in a position perpendicular to the stripe momentum is indistinguishable from the scenario where the pseudo-gap is due to the superconductivity in the stripes.

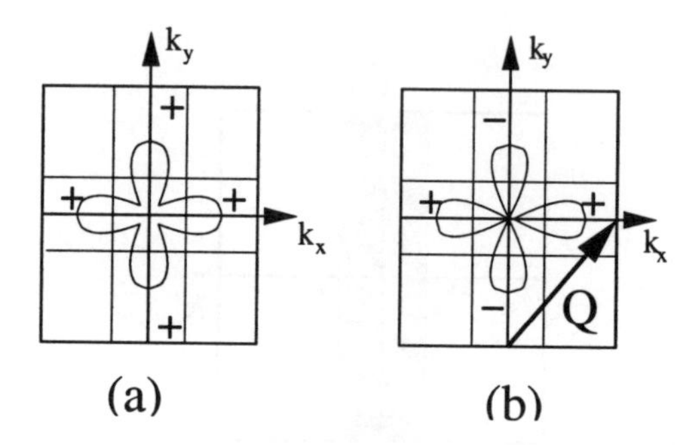

Figure 4. Possible pairing states for a system with various stripe domains.

Notice that each pseudo-gap can be associated with the stripes that are perpendicular to them. Furthermore, by generating samples where one type of domain is preferred (by growing the samples under pressure, for instance [32]) the d-wave state becomes anisotropic and eventually turns into an s-wave state if only one kind of domain exists. An interesting consequence of this kind of anisotropy generated by the domains is related with the amplitude of the pseudo-gap relative to the orientation of the stripes, namely, the pseudo-gap should be stronger in the direction perpendicular to the dominating domain. This effect makes possible the distinction between theories that assume the stripes to be superconducting and ours. Since the size of the pseudo-gap is proportional to the superfluid stiffness [19], a system with unbalanced of domains will be an anisotropic superconductor in the plane with an anisotropic superfluid stiffness. In a Ginzburg-Landau theory it implies three different different penetration depths and therefore in the presence of a magnetic field perpendicular to the planes (along the c-axis) the superconducting vortices acquire elliptical shape [29]. The longer axis of the vortex is the one with larger superfluid stiffness and therefore in our theory it should be perpendicular to the stripe orientation (see Fig.5(a)). On the other hand, if the pseudo-gap is a result of superconductivity coming from the stripes alone the larger axis should be parallel to the stripe orientation like in Fig.5(b). The distinction between these two orientations can be observe in scanning tunneling microscope (STM) experiments of single vortices in mono-domain regions of the crystal [30].

For the π-junction mechanism to work one needs to couple domains in some particular way so that not only one gets a coherence between different

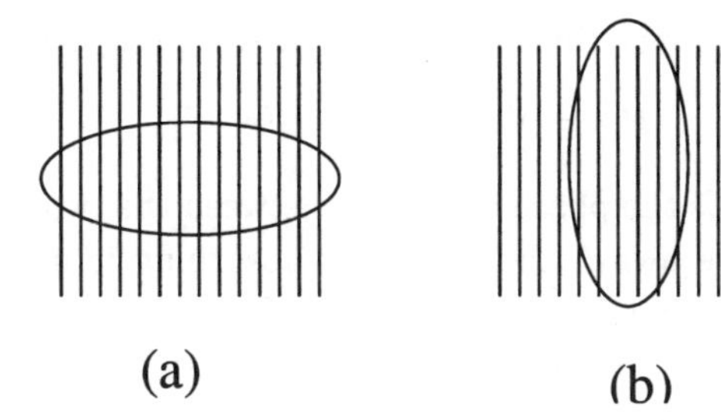

Figure 5. Straight lines represent the direction of the stripes and the ellipsis the profile of a vortex in a magnetic field perpendicular to the plane.

domains but with a phase relationship that is the correct one, that is, the d-wave state should be preferred over the anisotropic s-wave. One should recall that usually π-junctions occur when Cooper pairs tunnel from two superconductors through a magnetic region [31]. The reason for the change of sign in the Cooper pair wavefunction is due to the strong on-site Coulomb repulsion that constrains the Cooper pair to be broken in a specific order before tunneling is achieved. Naturally, there is no lack of magnetic regions in these materials, especially in the under-doped regime where the magnetic response is observed in neutron scattering [32] and other magnetic resonance experiments [33]. Besides the need of a magnetic response one needs to have the correct wave-vector. It is clear that to couple two different pseudo-gap regions one needs a wave-vector $\mathbf{Q}_M = (\pm\pi/a, \pm\pi/a)$ that links different domains as shown in Fig.4(b).

In order to simplify the treatment of the problem let us assume an infinitely long interface between two domains [27]. Let us name the bosonic operator that creates (annihilates) this magnetic mode R ($R^\dagger$), and the creation (annihilation) operators for the bosonic states residing in the two pseudo-gap regions $(-\pi/a, 0)$ and $(0, \pi/a)$ will be B_v ($B_v^\dagger$) and B_h ($B_h^\dagger$), respectively. The coupling between different domains is such that one boson in one domains absorbs a magnetic mode and the pair is transferred to the other domain, that is, the coupling is of the form:

$$H_D = \lambda(RB_h^\dagger B_v + h.c.) \tag{4}$$

where $\lambda > 0$ is the coupling constant. The action that describes this problem can be written as:

$$S = \int d\tau \int d^2x \left(\bar{R}(\partial_\tau + E_R(\nabla))R + \sum_{\alpha=h,v} \bar{B}_\alpha(\partial_\tau + E_B(\nabla))B_\alpha + H_D \right), \tag{5}$$

where τ is the imaginary time. Using (2) the coupling can be written as:

$$H_D = \lambda|\Delta|(Re^{i\varphi} + R^\dagger e^{-i\varphi}) \tag{6}$$

where $\varphi = \varphi_v - \varphi_h$ is the phase difference between the two domains. Notice that we can "gauge away" the phase by absorbing it into the definition of the operator R, that is, we can define $\tilde{R} = Re^{-i\varphi}$. In this case the Hamiltonian (6) is a "source" term in the problem and therefore it has a finite expectation value in the ground state, namely, $\langle R \rangle \neq 0$, that is the boson R condenses at low temperatures. In this case the energy of the problem is given by:

$$E_D = \lambda|\Delta||\langle R \rangle| \cos(\varphi) \tag{7}$$

which is minimized for $\varphi = \pi$ leading to π-junction and therefore to d-wave superconductivity.

Because of the symmetry of the problem it is clear that φ can only depend on τ and z, the coordinate perpendicular to the interface. Assuming that φ is varies smoothly in space we can obtain the phase action by tracing over the R modes:

$$S_{eff} = \frac{\rho_s}{2} \int d\tau \int dz \left[(\partial_z \varphi)^2 + c_s^{-2} (\partial_\tau \varphi)^2 \right] \tag{8}$$

where $\rho_s \propto \lambda^2|\Delta|/\omega_R^2$ is the phase stiffness (ω_R is the characteristic frequency of the R modes) and c_s is the superfluid velocity (from now on we set $c_s = 1$). Notice that (8) describes a 2D XY model which has power law correlations for distances smaller than the correlation length ξ:

$$\langle e^{i\varphi(r)} e^{-i\varphi(0)} \rangle \propto (\Lambda r)^{-\eta} \tag{9}$$

where $r = \sqrt{z^2 + \tau^2}$, $\eta = 1/(2\pi\rho_s)$ and Λ is a large momentum cut-off. For $r > \xi$ the above correlation function vanishes. The system has a Kosterlitz-Thouless transition to quasi-long order ($\xi \to \infty$) when $\rho_{s,c} = 2/\pi$ [34]. Since the calculation here was done at zero temperature it implies the existence of a quantum critical point as a function of ρ_s.

One can also study the effect of the quasi-long range order in the physics of the magnetic mode at $\mathbf{Q}_M$. Consider, for instance the correlation function for the R mode for $r < \xi$:

$$\langle R^\dagger(z,\tau) R(0,0) \rangle = \langle e^{-i\varphi(r)} \tilde{R}^\dagger(r) \tilde{R}(0) e^{i\varphi(0)} \rangle$$
$$\approx \langle \tilde{R}^\dagger(z,\tau) \tilde{R}(0) \rangle \langle e^{-i\varphi(r)} e^{-i\varphi(0)} \rangle \approx e^{-i\omega_R \tau} (\Lambda r)^{-\eta} \tag{10}$$

where in the second line we used a perturbative decoupling and assumed that the R modes are essentially dispersionless. It is interesting to consider the Fourier transform of the correlation function:

$$S(\omega,k)=\int_{-\infty}^{+\infty}d\tau\int_{-\infty}^{+\infty}dz\,e^{i(\omega\tau-kz)}\,\langle R^{\dagger}(z,\tau)R(0,0)\rangle \tag{11}$$

where k measures the momentum from $\mathbf{Q}_M$. Using (10) we see that:

$$S(\omega,0)\propto\frac{\cos\left[(2-\eta)\arctan((\omega-\omega_R)\xi)\right]}{(\xi^{-2}+\omega^2)^{1-\eta/2}}\,. \tag{12}$$

Notice that in the ordered phase $\xi\to\infty$ the magnetic mode becomes infinitely sharp at $\omega=\omega_R$. However, in the disordered phase the mode acquires a width proportional to ξ and its intensity behaves like $\xi^{2-\eta}$. The behavior of the magnetic mode described here is very similar to the so-called resonance mode at 41 meV observed in YBCO [35] at $(\pi/a,\pi/a)$ that becomes very sharp inside of the superconducting phase but disappears in the normal phase.

4. Conclusions

We have presented here a possible theoretical description for the development of superconducting correlations in stripe cuprates. We have argued that the stripe system is not really one-dimensional but in fact quasi-two-dimensional. We argue that in the presence of electron-electron interactions alone the CuO_2 planes are insulating and that superconductivity is only possible in the presence of optical phonons that are responsible for the hybridization between stripe states and d-wave pairs. As a consequence of a non-trivial electron phonon coupling one expects lattice distortions to be directly coupled to superconducting fluctuations (vortices) and that the process of phase coherence requires more than one topological excitation. Moreover, we have argued that most of the action in the superconducting state occurs perpendicular to the stripes, instead of parallel to them, and that this behavior can be tested experimentally. In particular, we have shown that the pseudo-gap develops in the the Brillouin zone in the direction transverse to the stripe momentum and show that phonon anomalies should appear in the same regions where the pseudo-gap opens.

We also have shown that for a mono-domain stripe array the superconducting state is of the anisotropic s-wave type and that a d-wave state is only possible when orthogonal domains of stripes are coupled to each other by magnetic excitations with momentum $(\pi/a,\pi/a)$. We have shown that while the magnetic excitations with this momentum are very broad in the normal phase they became very sharp when the domain acquire phase coherence through a Kosterlitz-Thouless like transition. This behavior is reminiscent of the resonance peak observed in YBCO.

Acknowledgments

I would like to acknowledge E. Carlson for a collaboration in the problem of resonance peak and for many illuminating discussions, and Sasha Chernyshev for important comments.

References

1. G. Bednorz and K. A. Müller, Z. Phys.B: Condens. Matter **64**, 189 (1986).
2. N. F. Mott in *Metal-Insulator Transitions* (Taylor & Francis, London, 1974).
3. S. Chakravarty *et al.*, Phys. Rev. B **39**, 2344 (1989)
4. T. D. Stanescu and P. Phillips, unpublished.
5. P. W. Anderson, Science **235**, 1196 (1987).
6. S. R. White and D. J. Scalapino, Phys. Rev. Lett. **80**, 1272 (1998).
7. J. M. Tranquada *et al.*, Nature, **375**, 561 (1995).
8. S. R. White and D. J. Scalapino, Phys. Rev. B **60**, 753 (1999).
9. B. P. Stojkovic *et al.*, Phys. Rev. B **62**, 4353 (2000).
10. B. Normand, and A. P. Kampf, Phys. Rev. B **64**, 024521 (2001).
11. See, E. Dagotto, Rev. Mod. Phys. **66**, 763 (1994), and references therein.
12. S. Trugman, Phys. Rev. B **37**, 1597 (1988).
13. G. B. Martins *et al.*, Phys. Rev. Lett. **84**, 5844 (2000).
14. A. L. Chernyshev *et al.*, Phys. Rev. Lett **84**, 4922-4925 (2000).
15. A. L. Chernyshev *et al.*, Phys. Rev. B **65**, 214527 (2002); G. B. Martins *et al.*, Phys. Rev. B **62**, 13926 (2000).
16. J. Zaanen, and O. Gunnarsson, Phys. Rev. B **40**, 7391 (1989); D. Poilblanc, and T. M. Rice, Phys. Rev. B **39**, 9749 (1989); H. J. Schulz, J. Phys. (Paris) **50**, 2833 (1989); K. Machida, Physica C **158**, 192 (1989).
17. A. H. Castro Neto, Z. Phys. B-Cond. Matter, **103**, 185 (1997).
18. N. Hasselmann *et al.*, Phys. Rev. Lett. **82**, 2135-2138 (1999).
19. A.H.Castro Neto, Phys. Rev.B **64**, 104509 (2001).
20. C. Buhler *et al.*, Phys. Rev. Lett. **94**, 2690 (2000).
21. V. J. Emery *et al.*, Phys. Rev. B **56**, 6120 (1997).
22. E. W. Carlson *et al.*, Phys. Rev. B **62**, 3422 (2000) .
23. D. S. Marshall *et al.*, Phys. Rev. Lett. **76**, 4841 (1996).
24. B. H. Toby *et al.*, Phys. Rev. Lett. **64**, 2414 (1990).
25. Z. X. Shen *et al.*, cond-mat/0108381.
26. Y. Wang *et al.*, Phys. Rev. B **64**, 224519 (2001).
27. This work is being developed in collaboration with E. Carlson.
28. A. H. Castro Neto and F. Guinea, Phys. Rev. Lett., **80**, 4040 (1998).
29. E. Carlson *et al.*, unpublished.
30. J. E. Hoffman *et al.*, Science **295**, 466 (2002); C. Howald *et al.*, cond-mat/0201546.
31. B. Spivak and S. Kivelson, Phys. Rev. B **43**, 3740 (1991).
32. H. A. Mook *et al.*, Nature **404**, 729 (2000).
33. M.-H. Julien *et al.*, Phys. Rev. Lett. **84**, 3422 (2000).
34. J. M. Kosterlitz and D. J. Thouless, J. Phys. C **6**, 1181 (1973).
35. J. Rossat-Mignod *et al.*, Physical C **185-189**, 86 (1991).

ON THE CHARACTERISTIC LOCAL LATTICE DISPLACEMENTS IN THE HIGH T_c OXIDES

N.L. SAINI, A. BIANCONI

Unitá INFM and Dipartimento di Fisica, Università di Roma "La Sapienza" P.le Aldo Moro 2, 00185 Roma, Italy

H. OYANAGI

National Institute of Advanced Industrial Science and Technology, Tsukuba Central 2, 1-1-4 Umezono, Tsukuba, Ibaraki 305, Japan

We have studied local structure of the high T_c superconducting copper oxides to explore a possible relationship between the characteristic atomic displacements, superconductivity and the charge inhomogeneous state of these complex oxides. High k-resolution polarized Cu K-edge extended x-ray absorption fine structure (EXAFS) has been used to determine the characteristic CuO_2 local displacements with variable disorder and chemical pressure. Temperature dependent local atomic displacements show anomalies, at the T_c and a temperature T_s where the charge inhomogeneous state appears, as revealed by a change in the correlated Debye-Waller factor of the Cu-O bonds. While the Debye Waller factor shows a clear drop at the T_c, an order parameter like change appears at the T_s. We also report local structure of the Nb_3Ge intermetallic superconductor by Ge K-edge EXAFS with an emphasis to determine the local and instantaneous displacements across the T_c. We find similar change in the correlated Debye-Waller factor of the Ge-Nb bonds showing a drop at the T_c. Analogous local atomic displacements in this short coherence length superconductor indicates importance of the local electron-lattice interactions not only for their superconductivity but also for the normal state properties.

1. Introduction

Understanding various kind of competing ground states of the complex materials with strong electron-electron correlation is one of the key tasks in the condensed matter physics. Transition metal oxides with perovskite structure, which represent a family of such complex materials, have stimulated considerable scientific and technological interest in the recent past because of their exotic properties and possibility of tailoring new materials with desired properties. In fact, the oxides possess a variety of novel properties such as high-temperature superconductivity (HTS), colossal magneto-resistance (CMR), metal-insulator transitions (MIT) etc. In addition, these oxides manifest self-organization of various degrees of freedom (stripes) at a mesoscopic length-scale, the phenomena which has been a point of recent debate in the field [1].

Structurally the case of these oxides is rather non-trivial and knowledge of long-range crystallographic structure is not enough to explain their basic properties unlike the simple solids such as normal metals. Indeed, the basic characteristics of these doped oxides depend strongly on the local atomic structure, as revealed by a series of experiments. In fact, the elastic fields due to local strain play vital role in the physics of these complex oxides [2, 3]. Moreover, the situation seems quite close to what appears in some martensitic systems where local structural tweeds and twins have direct implication on their basic electromic character [4].

Let us take the example of cuprates showing the high T_c superconductivity, which are layered materials with CuO_2 planes, separated by insulating rocksalt oxide layers. The importance of the CuO_2 plane has created major interest to study the electronic versus structural behavior of this structural unit. Even if the fundamental character of the superconducting order parameter with a charge 2e remains intact, understanding of the mechanism is stagnated by interplaying low temperature orders, related with the charge, spin and lattice degrees of freedom, that can compete or coexist with the superconductivity. In fact, the pending problem is the nature of the coupling mechanism responsible for creating the pairs in the CuO_2 plane. Recent experiments support a key role of local electron-lattice interactions [5-11]. At this point it becomes important to distinguish and quantify the characteristic local lattice distortions associated with the fundamental electronic character of these materials, either the superconductivity or the normal state behavior, as the quantitative value of the displacements could distinguish proper model based on elactron-lattice interactions [12]. As a matter of fact, there has been large ambiguity in the measured magnitude of the atomic displacements, depending on the time scale of the experimental technique used, due to small disorder. However, recent advances in the materials growth and development of new experimental techniques at higher experimental facilities (synchrotron radiation sources and neutron sources) have geared up our understanding to the intrinsic structural physics in these HTS copper oxides.

The main experimental probes used to determine the local displacements in the complex oxides are the pair distribution function (PDF) analysis of neutron and x-ray diffraction, extended x-ray absorption fine structure (EXAFS) and ion channeling [5-10]. All these techniques have their own limitations to provide information on the quantitative atomic displacements, however, the results on the local lattice displacements determined by these techniques agree quite well. The EXAFS spectroscopy, a fast (~10^{-15} sec) and local (~5-6Å) tool [13], has been widely applied to the study of the complex oxides during the past decade. Availability of the high brilliance and polarized x-ray synchrotron radiation sources has been an added advantage for the technique allowing quantitative determination of the directional atomic displacements around a selective site in complex systems such as HTS cuprates [9, 10, 14-16] and CMR manganites [17]. In fact, with recent technical advances, the EXAFS spectroscopy offers unique approach to pin point short-range atomic displacements and their dynamics.

This paper is aimed to provide determination of the local atomic displacements in the short coherence length HTS oxides and related systems to explore possible implication of these atomic displacements in their basic properties. We will focus on the problems of intrinsic local charge and structural inhomogenieties and superconductivity in the copper oxides. We will also report some new results on intermetallic compounds showing short coherence length superconductivity to point out similarities and differences between the two classes of materials. Here we have exploited the Cu K-edge EXAFS with high k-resolution to study the local atomic displacements in the electronically active Cu-O networks in the HTS copper oxides. The temperature dependent distribution of the local lattice distortions (dynamic and static) are measured by the correlated Debye-Waller factor of the Cu-O bonds (σ^2_{CuO}) revealing anomalous change across the charge stripe ordering temperature. We find a clear change in the charge stripe ordering temperature with variation of the disorder and the chemical pressure, showing the upturn with different amplitude. In addition, σ^2_{CuO} at the superconducting T_c shows a drop with variable amplitude as a function of the chemical pressure. A similar change is found in the Debye-Waller factor of the Ge-Nb bonds in the Nb_3Ge intermetallic compound suggesting a clear intimacy between the local atomic displacements and the short coherence length superconductivity.

2. Experimental details

The Cu K-edge x-ray absorption measurements on well-characterized superconducting single crystals of different superconducting systems were performed at the beamlines BM29 and BM32 at the European Synchrotron Radiation Facility (ESRF), Grenoble and BL13B of Photon Factory, Tsukuba. At the BM29 the synchrotron radiation emitted by a Bending magnet source at the 6 GeV ESRF storage ring was monochromatized by a double crystal Si(311) monochromator. For temperature dependent measurements at the BM29 the samples were mounted in a closed cycle two stage He cryostat. Fluorescence yield (FY) off the samples was collected using 13 Ge element solid state detector to measure the absorption signal at the BM29. A Si(111) crystal was used as monochromator and a 30-element Ge x-ray detector array was used to measure the absorption spectra at the BM32. At the BL13B the synchrotron radiation emitted by a 27-pole wiggler source at the 2.5 GeV Photon

Factory storage ring was monochromatized by a double crystal Si(111) and sagittally focused on the sample. For the Nb_3Ge system, temperature dependent Ge K-edge absorption measurements were performed on well characterized thin films, grown by the chemical vapour deposition (CVD) showing sharp superconducting transition below 20.6 K. The spectra were recorded by detecting the fluorescence photons using a 19-element Ge x-ray detector array. The sample temperatures were controlled and monitored to within an accuracy of ±1 K. As our standard experimental approach, several absorption scans were collected to limit the noise level to the order of 10^{-4}. Standard procedure was used to extract the EXAFS signal from the absorption spectrum [13] and corrected for the x-ray fluorescence self-absorption before the analysis. Further details on the experiments and data analysis could be found in our earlier publications [14-17].

3. Results and discussion

3.1 Disorder effect on the local lattice displacements and the charge stripes ordering in the HTS La-based copper oxides

Let us start with one of the popular issues in the copper oxides, the charge stripe ordering. Indeed this new approach based on inhomogeneous charge distribution in stripes is making its ground among conventional theories to understand the physics of these superconductors. Several experiments probing different characteristics related with charge, spin and lattice dynamics are found to be consistent with the striped phases in these materials [5-10, 14-16, 18-29]. The La_2CuO_4 family of copper oxide superconductors has been one of the most popular ones to uncover the inhomogeneous charge transport and charge stripe ordering in the HTS copper oxides by variety of experimental techniques [10, 18-28]. A prototype example is the $La_{1.48}Sr_{0.12}Nd_{0.4}CuO_4$ system [18, 20, 22, 24, 28] in which the static charge stripe order has been observed. Here we will address the effect of disorder on the charge stripe order and the associated local lattice displacements in the CuO_2 plane. We have studied $La_2CuO_{4.1}$ (LCO), $La_{1.85}Sr_{0.15}CuO_4$ (LSCO) and $La_{1.48}Nd_{0.4}Sr_{0.12}CuO_4$ (LNSC) systems representing respectively the oxygen doped, Sr doped and Nd doped systems with increasing disorder. Temperature dependent distribution of the local lattice distortions (dynamic and static) is measured by the correlated Debye-Waller factor of the Cu-O bonds using high resolution in-plane polarized Cu K-edge EXAFS.

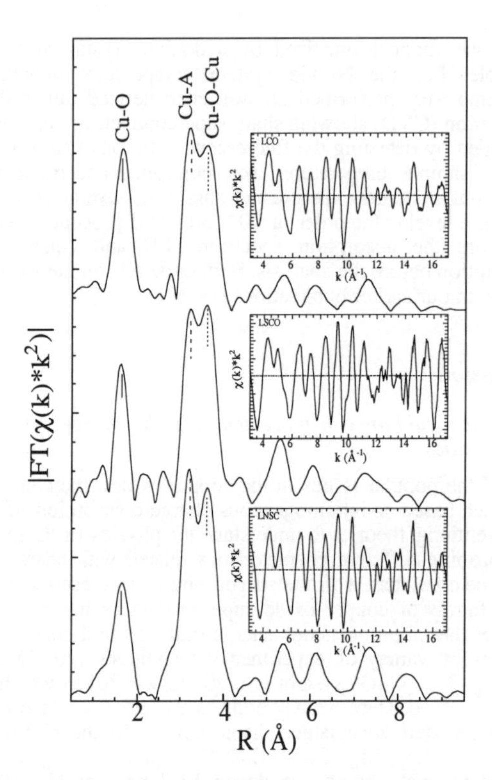

Figure 1. Fourier tranforms of the EXAFS oscillations (multiplied by k^2) of LCO (upper), LSCO (middle) and LNSC (lower) systems showing global atomic distribution around the Cu in the three systems. The Fourier transforms are performed between k=3-17 $Å^{-1}$ and not corrected for the phase shifts. The respective EXAFS oscillations are shown as insets.

Fig. 1 shows Fourier transforms (FT) of the representative polarized Cu K-edge EXAFS of LCO, LSCO and LNSC crystals measured at low temperature (20 K) with E vector of the plane polarized x-rays falling parallel to the CuO_2 square plane. The EXAFS plots (shown as insets) represent the raw data and are shown multiplied by k^2 to emphasize the higher k-region. The FT provide a global atomic distribution around the absorbing Cu atom in the measured systems and peaks appear due to scattering of the photoelectron, ejected at the Cu site, with the near neighbor atom. The main peaks in the FT are denoted by Cu-O, Cu-A (A=La, (La, Sr) and (La, Nd, Sr) respectively for the LCO, LSCO and LNSC systems) and Cu-O-Cu, appearing due to scattering of the ejected photoelectron at the Cu site with the nearest in-plane oxygen atoms (at ~1.9 Å), A atoms (sitting at ~3.2Å and 45° from the direction of the photoelectron) and the next Cu atom (at ~ 3.8Å), respectively. There are evident differences in the FT measured on the three systems, mainly around the Cu-A peak due to different contributions. The Cu-A peak is largely damped for the case of LCO in compare to the one for the LSCO and LNSC systems. There is no reason for this damping except the fact that the doped oxygen in this system occupies the space between the La sites, introducing a larger disorder in the La position with respect to the central Cu [30]. The absolute differences may be difficult to extract because of complex interference effects due to different origins of the backscattering in the three systems at the Cu-A position and interference with the Cu-O-Cu multiple scattering.

Here we focus our attention on the atomic displacements in the electronically active CuO_2 plane (i.e. the in-plane Cu-O bond). In the in-plane polarized Cu K-edge EXAFS the signal due to the Cu-O bond distances is well separated from the longer bond contributions and can be easily extracted and analyzed separately. In this work we have used the 'standard procedure' for the analysis of EXAFS data considering a single distance for the coordination shell, where the effective Debye-Waller factor (DWF) includes all distortion effects, taking into account both static and dynamic distortions. This standard approach is adopted to make a direct comparison of the temperature dependent distortions in different systems, where the correlated DWF of the Cu-O pairs, σ^2_{CuO} is a suitable order parameter of the local CuO_2 distortions. Quantitative value of the σ^2_{CuO} depends on technical aspects (experimental geometry and analysis), however, this is irrelevant for the temperature dependence. Within the reported uncertainties, we have ensured the quantitative values for the DWF in different systems by measuring the three systems in same experimental conditions and applying the same data analysis procedure.

We have determined the σ^2_{CuO} by modeling the Cu-O EXAFS considering a single Cu-O bond, as revealed by diffraction measurements. The Cu-O EXAFS was simulated in the same k ($k=3\text{-}17Å^{-1}$) range for all the systems. The number of parameters which may be determined by EXAFS is limited by the number of independent data points: $N_{ind} \sim (2\Delta k\Delta R)/\pi$, where Δk and ΔR are respectively the ranges in k and R space over which the data are analyzed. In the present case $\Delta k=14$ $Å^{-1}$ and $\Delta R=1$ Å give $N_{ind}\sim 9$ for the first shell EXAFS. Except the radial distance R and the σ^2_{CuO}, all other parameters were kept constant in the conventional least squares paradigm following the standard approach and our experience on the similar systems [9, 10, 14-17]. The average distances were independent of temperature and found to be similar to the one determined by the diffraction experiments on the three systems.

Temperature dependence of σ^2_{CuO} is shown in Fig. 2. It is known that at the appearance of any charge density wave like instability the Debye-Waller factor shows an anomalous change as found in several density wave systems [31]. Indeed the temperature dependence of the σ^2_{CuO} shows an anomalous upturn at a temperature T_s, due to the instability (driven by a particular local lattice distortion in the CuO_2 plane [10, 14]). Considering evidences of charge stripe ordering in the model LNSC system appearing at ~60 K, provided by several experimental techniques [18, 19], we assign the anomalous upturn in the σ^2_{CuO} to a charge instability giving charge stripe ordering. The results are also consistent with the charge stripe ordering in the LCO system below ~190 K, revealed by x-ray diffraction [32]. We could also see that the increase of σ^2_{CuO} below the temperature T_s is followed by a decrease around the superconducting transition temperature T_c for the superconducting LCO and LSCO systems, however, the LNSC does not show such a change, as the system remains non-superconducting down to the temperature measured.

Effect of disorder on the local structure and the charge stripe ordering is clear in Fig. 2. The anomalous increase in σ^2_{CuO} due to stripe-charge ordering appears at different temperature in the different systems, below ~190 K, ~100 K and ~60 K in the LCO, LSCO and LNSC systems respectively, suggesting that the disorder in the CuO_2 plane brings down the charge stripe ordering temperature. In addition, an increase in σ^2_{CuO} above the charge stripe order temperature for LNSC is observed that provides a clear indication of increased microscopic disorder in the CuO_2 plane. As a matter of fact, the dynamic contribution to the lattice fluctuations is decreased, as evidenced from the step-like increase with smaller amplitude across the charge stripe order temperature T_s in the LNSC system in which the static disorder dominates.

Here we summarize this section, we have studied effect of disorder on the local lattice displacements and charge inhomogeneities in the La_2CuO_4 family of superconducting copper oxides. The correlated Debye Waller factor of the Cu-O pairs, measuring static and dynamic displacements in the CuO_2 plane shows an abrupt change with an order parameter like behavior across the stripe ordering temperature in the LCO, LSCO and LNSC systems. We find that the charge stripe ordering temperature gets reduced with extrinsic disorder. The anomalous step-like upturn appears with smaller amplitude in the non-superconducting LNSC in which more static charge stripes have been found. Furthermore, we find that the appearance of the superconducting state is accompanied by a decrease of the instantaneous local lattice distortions providing an evidence of the role of electron-lattice interactions to be an important ingredient for the superconducting pairing.

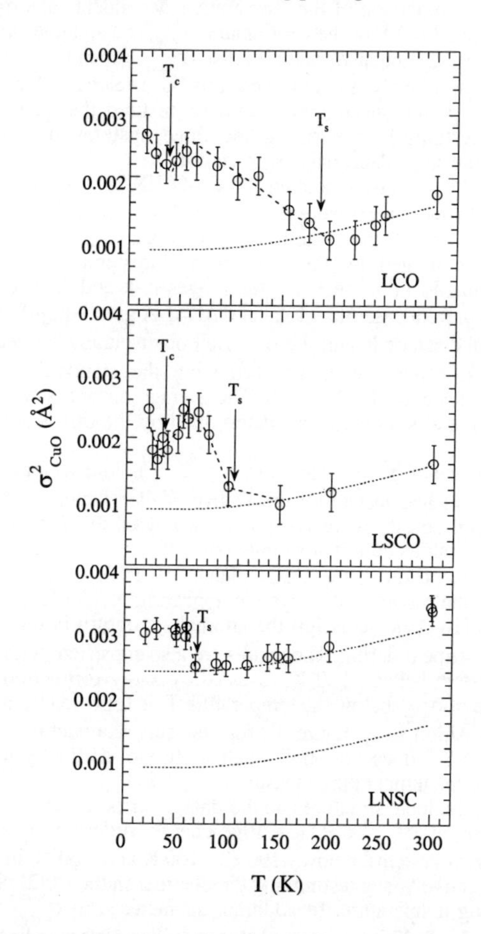

Figure 2. Temperature dependence of the correlated Debye-Waller factors (symbols) of the Cu-O pairs (σ^2_{CuO}) for the LCO (upper), LSCO(middle) and LNSC(lower) systems. The expected temperature dependence of the Debye-Waller factor for a fully correlated motion of Cu and O is shown by a dotted line. The dashed line across the charge stripe order temperature (T_s), revealed by the anomalous upturn, is for guide to the eyes. A constant value of 0.00145 is added to guide the temperature dependence of the experimental Debye-Waller factor for the LNSC system (lower).

3.2 Local lattice displacements as a function of chemical pressure on the CuO_2 plane in the HTS copper oxides

Structurally the HTS copper oxides are heterogeneous materials made of alternated layers of metallic body centered cubic (bcc) CuO_2 layers and insulating rock-salt face centered cubic (fcc) M-O (M= Ba, Sr, La) layers [33, 34]. The mismatch between the two sub-lattices is generally estimated by $1\text{-}t=[r(A\text{-}O)]/\sqrt{2}[r(Cu\text{-}O)]$ where r(A-O) (i.e., r(La-O), r(Sr-O) and r(Ba-O)) and r(Cu-O) are the respective bond lengths and t is the Goldschmidt tolerance factor [34]. Due to the lattice mismatch the CuO_2 sheets are under compression and (M-O) layers under tension. Here we have investigated local atomic displacements as a function of the micro-strain in the CuO_2 plane due to the lattice mismatch. The $La_2CuO_{4.1}$ (LCO) with T_c~40 K, $Bi_2Sr_2CaCu_2O_{8+\delta}$ (Bi2212) with T_c~87 K and $HgBa_2CuO_{4+\delta}$ (Hg1201) with T_c~94 K (and $HgBa_2CaCu_2O_{6+\delta}$ (Hg1212) with T_c~116 K) systems are used as representatives for the La-based, Bi-based and Hg-based superconducting families containing respectively the La-O, Sr-O and Ba-O as rock-salt layers that sustain different chemical pressure on the CuO_2 planes, where the dopants are interstitial oxygen ions in the block layers.

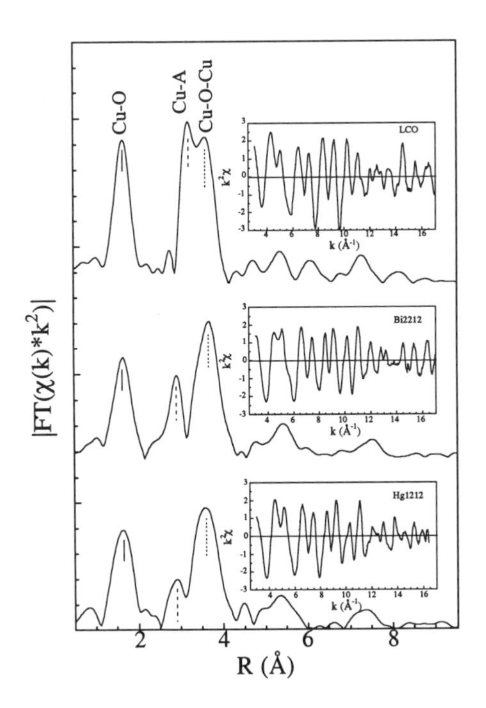

Figure 3. Fourier transform of the Cu K-edge EXAFS measured on LCO (upper), Bi2212 (middle) and Hg1212 (lower). The inset shows the corresponding EXAFS oscillations.

Fig. 3 shows Fourier transforms (FT) of the Cu K-edge EXAFS measured on different superconducting systems in the in-plane geometry. The insets show respective EXAFS oscillations. There are evident differences in the FT of the EXAFS spectra measured on different systems. The major differences appear around the Cu-M (M=La, Sr(Ca), Ba(Ca)) peak due to different block-layers. Again, the Cu-O Debye Waller factor (σ^2_{CuO}), is the parameter used to make a systematic comparison between the systems with variable microstrain. The temperature dependence of the σ^2_{CuO} is shown in Fig. 4 for the LCO (T_c~40 K), Bi2212 (T_c~87K) and Hg1201 (T_c~94K) systems representing the three different families of the HTS copper oxides.

From the temperature dependence of σ^2_{CuO} we can easily define at least two anomalous temperatures where the σ^2_{CuO} show anomalies. There is an anomalous increase below the stripe ordering temperature T_s followed by a decrease around the superconducting transition temperature T_c. The increase at T_s appears in the LCO and Bi2212 systems, however, the Hg1201 system does not show any evident up turn. On the other hand, the drop in σ^2_{CuO} at the superconducting transition temperature T_c appears common to all the systems (however, less evident in the LCO system). The Hg1201 system manifests a large decrease in the σ^2_{CuO} around the superconducting transition temperature. Here we should mention that, apart from the static and dynamic distortions of the CuO_2 lattice, σ^2_{CuO} contains contribution from the thermal vibrations. However, in the present case the thermal contribution to σ^2_{CuO} should be similar for all the systems and hardly affects the present discussion.

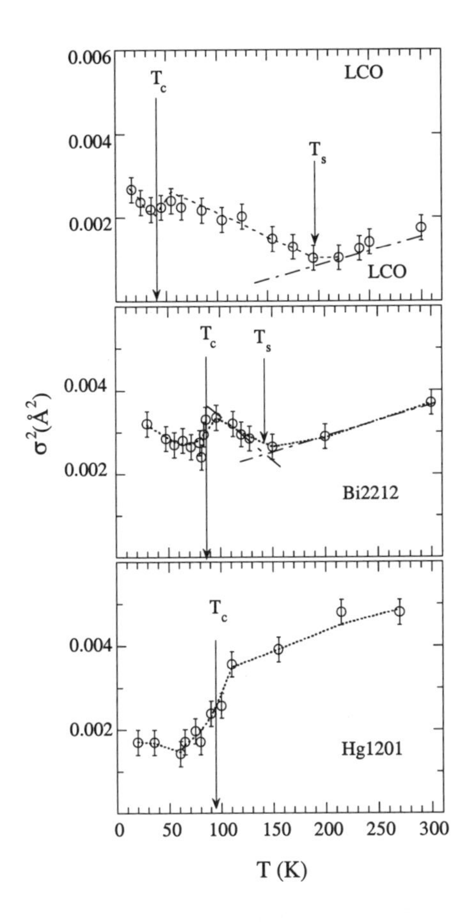

Figure 4. Temperature dependence of the Cu-O Debye-Waller factor σ^2 determined by EXAFS; $La_2CuO_{4.1}$ (upper), Bi2212 (middle) and Hg1201(lower). The dashed line is a guide to the eyes. The resulting Debye-Waller factor σ^2 shows abnormal temperature dependence with an increase below a temperature T_s followed by a decrease around the superconducting transition temperature T_c. The error bars represent the average estimated noise level.

As discussed above, the anomalous increase in σ^2_{CuO} is due to stripe ordering in the CuO_2 plane of the LCO (T_s~190 K) and Bi2212 (T_s~140 K) systems. Recently Sharma et al [6] have further confirmed our earlier results and found a clear upturn in the temperature dependence of the excess displacements (a parameter similar to the Debye-Waller factor measuring dynamic and static distortions) measured by ion-channeling on the YBCO system at the stripe ordering temperature. In fact, below this temperature the pair distribution function becomes larger than that due to thermal fluctuations and the formation of striped phase should give an asymmetric PDF due to splitting of the Cu-O bonds as demonstrated earlier [9, 10].

The lower temperature anomaly in σ^2_{CuO} appears around the superconducting transition temperature T_c. The correlated Debye-Waller factor σ^2_{CuO} shows an anomalous decrease around the T_c. This is a clear indication that the appearance of the superconducting state is accompanied by a decrease of the instantaneous local atomic displacements pointing towards a key role of local electron lattice interactions in the superconducting pairing. Interestingly, the drop at the T_c is different for different systems and found to be maximum for the Hg-based compound where the block-layers are Ba-O with smaller micro-strain in the CuO_2 plane than the case of Bi2212 (Sr-O) and LCO (La-O). Below the T_c the σ^2_{CuO} shows a small increase, but this is within the limits of experimental uncertainties. At the superconducting transition the drop of the σ^2_{CuO} revealing decreased distortions at T_c could be due to decrease of the incoherent local lattice distortions to transfer electron lattice interaction energy in the pairing mechanism entering into a coherent state.

3.3 Signature of local structure anomaly at T_c in the Nb_3Ge superconductor

Here we wish to address the question whether the change in the local atomic displacements across the superconducting transition is the sole property of the HTS copper oxides or is a generic feature of short coherence length superconductors. We have taken Nb_3Ge intermetallic superconductor as an example and studied the local atomic displacements by Ge K-edge EXAFS measurements, with an emphasis to determine the local and instantaneous atomic displacements across the superconducting transition temperature T_c.

It should be recalled that until the discovery of the high T_c superconductivity in the copper oxides, the superconducting intermetallic compounds with A15 type structure were called high T_c superconductors. The highest T_c was found in Nb_3Ge compound belonging to the A15 structural family. These superconductors were widely studied for their characteristic properties, well reviewed by J. Muller [35]. The structural aspects of these systems are focussed in the review by Testardi [36]. There were some efforts to study the local structure of this class of superconductors by EXAFS measurements [37-40]. A highlight was the work by Cargill et al [39] who focussed on Nb K-edge EXAFS and revealed large anisotropic vibrational correlations related to the one dimensional Nb-Nb chains in the structure, which were much larger in the Nb_3Ge system than others A15 structures (Nb_3Al, V_3Si and Cr_3Si). Here we report evolution of the local structure of the Nb_3Ge in a temperature range of 6-300 K by high k-resolution fluorescence EXAFS, focussing in the region of superconducting transition temperature to explore possible local displacements. The Ge K-edge EXAFS reveals signature of a clear change in the correlated Debye-Waller factor of the Ge-Nb bonds at the superconducting transition temperature T_c, similar to the case of HTS copper oxides.

The Fourier transforms (FT) of the Ge K-edge EXAFS spectra (multiplied by k^2) at some representative temperatures are shown in Fig. 5. The Fourier transforms are performed between k=3-18 $Å^{-1}$ and not corrected for the phase shifts (representing the raw experimental data). An EXAFS spectrum is shown as an inset to the Fig. 5, showing oscillations up to k=18$Å^{-1}$. The main FT peak is denoted by Ge-Nb, representing the scattering of the ejected photoelectron at the Ge site with the nearest Nb atoms.

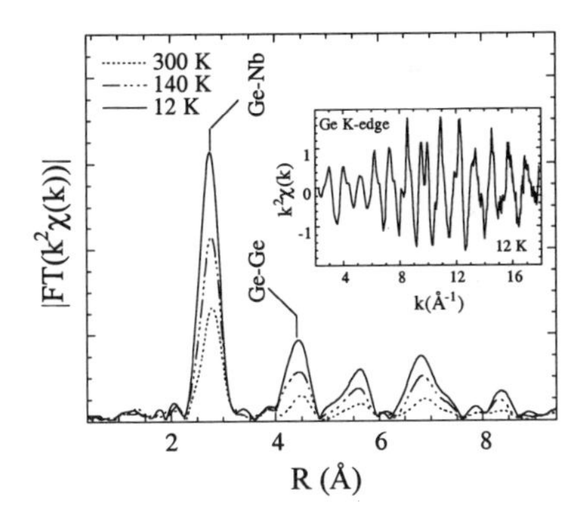

Figure 5. Fourier transforms (FT) of the EXAFS spectra (multiplied by k^2) recorded on the Nb_3Ge system. The FT have been performed between k=3-18 $Å^{-1}$ and not corrected for the phase shifts. The inset represents EXAFS oscillations.

Here we focus on the Ge-Nb displacements. We made an attempt to model the EXAFS by a single Ge-Nb distance following diffraction studies, however, unsatisfactory fit to the experimental data suggests another coexisting Ge-Nb distance. In fact a smaller Ge-Nb distance of ~2.66 Å was indicated to coexists with the crystallographic distance of ~2.87 Å [37]. Therefore, we used a starting model with two Ge-Nb distances, where the effective Debye-Waller factor (DWF) includes all distortion effects, taking into account both static and dynamic distortions. The starting parameters were taken from the diffraction studies on the system. In the present analysis, we have allowed to vary the two Ge-Nb bond lengths and their Debye Waller factors as a function of temperature. While the shorter Ge-Nb distance (~2.66 Å) shows a small but gradual increase with decreasing temperature, the longer bond length remains temperature independent ~2.88±0.005 Å, in the temperature range studied here, and hence this was fixed in the final iteration. Also, the relative probability of the two Ge-Nb bonds (32:68) was found to be temperature independent within the experimental uncertainties, and hence kept fixed for the final analysis.

Temperature dependence of the Debye-Waller factors of the Ge-Nb pairs, σ^2_{Nb1} and σ^2_{Nb2}, representing static and dynamic displacements around the Ge atoms, determined from the above analysis, are shown in Fig. 6. The DWF follows smooth behavior for both Ge-Nb bonds as a function of temperature, at least down to ~ 50K. Below this temperature there seems some divergence from this smooth behavior. The temperature dependence of the DWF for the two bonds could be fitted with an Einstein model of harmonic oscillator with Einstein temperatures θ_E=230°C and θ_E =210°C respectively for the σ^2_{Nb1} and σ^2_{Nb2} with constant values (0.003 and 0.0015 respectively) for the zero-point energies. While the value of θ_E =230°C for the σ^2_{Nb1} is consistent with the earlier reports [38, 39], we find distinct θ_E =210°C for the σ^2_{Nb2}.

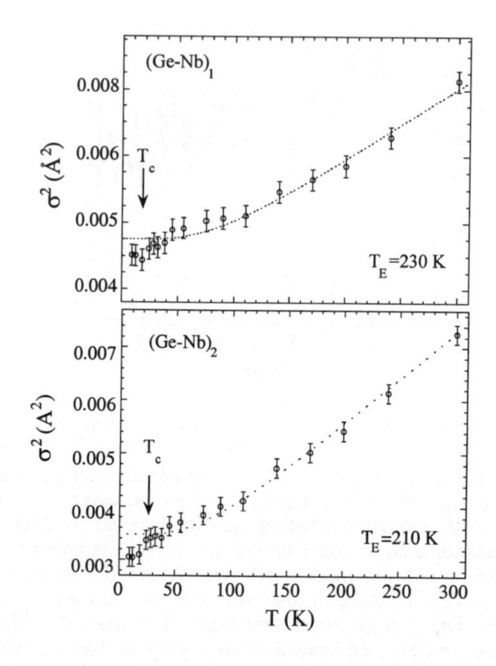

Figure 6. Temperature dependence of the correlated Debye-Waller factors (symbols) of the Ge-Nb pairs, σ^2_{Nb1} (upper) and σ^2_{Nb2} (lower). The expected temperature dependence of the Debye-Waller factors for a fully correlated motion of Ge and Nb, calculated by Einstein model, are shown by the dotted lines. The σ^2_{Nb1} and σ^2_{Nb2} are well described by the Einstein model with θ_E=230 K and 210 K respectively above ~ 50 K (see text).

What appears clear from the present data is that the divergence from the normal behavior of the DWFs, revealing anomalies in the local structure of the Nb_3Ge system. There seems an anomalous drop at the superconducting transition temperature T_c similar to the one observed in the high T_c cuprate superconductors by Cu K-edge EXAFS. The drop is much clear in the σ^2_{Nb2}. Interesting enough, the drop in the cuprates was found to depend on the critical temperature T_c, being larger while the T_c is higher. Fig. 7 shows the relationship between the superconducting transition temperature and the value of the drop in the DWF ($\Delta\sigma^2$) at the T_c for the cuprates, plotted with the drop in the DWF for the Nb_3Ge. While DWF for the Cu-O bonds are used, being the main electronic component, the DWF for the Ge-Nb is given considering the fact that the orthogonal Nb chains are the important structural elements for the electronic structure of the Nb_3Ge system. Interesting enough, the figure indicates that the two parameters, the T_c and $\Delta\sigma^2$ have proportionality like behavior. This result suggests an intimate relationship between the local lattice fluctuations and superconductivity in the short coherence superconductors having high transition temperature.

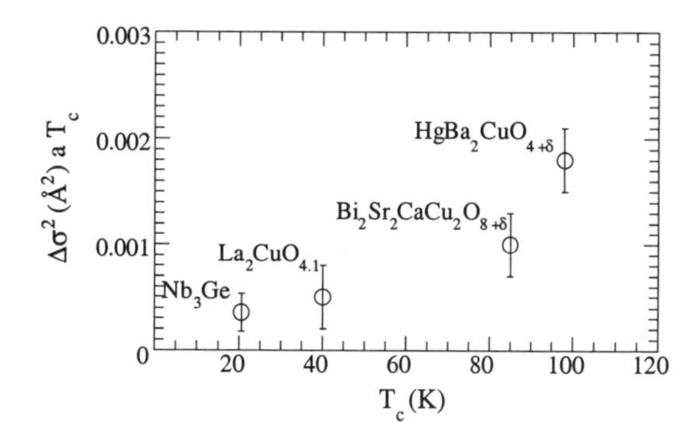

Figure 7. The drop in the Debye-Waller factors ($\Delta\sigma^2$) at the T_c for the HTS copper oxides is plotted as a function of the superconducting transition temperature, and compared with the drop for the Nb_3Ge.

In summary, we have measured local lattice displacements in the Nb_3Ge superconducting system. The temperature dependent local and instantaneous atomic displacements across the superconducting transition temperature, determined by the correlated Debye-Waller factor of the Ge-Nb bonds (σ^2_{Nb}), indicate a small but clear divergence from their normal temperature dependence. This change, appearing as a drop in the σ^2_{Nb} at the superconducting transition temperature T_c, is similar to the one found for the Debye-Waller factor of the Cu-O bonds, σ^2_{CuO} in the high T_c cuprate superconductors. The comparison of the drop at T_c in the Nb_3Ge intermetallic with the one observed in the cuprate superconductors reveals that the local displacements are tied to the superconductivity of the materials with small coherence length. Also, we find a second distance, shorter by ~0.2 Å from the average one, that seems to be related with the phase without any long-range crystallographic symmetry, indicating intrinsic inhomogeneous state of the Nb_3Ge system.

4. Concluding remarks

In conclusion, we have measured local lattice displacements in the short coherence length superconductors as perovskite copper oxides and intermetallic Nb_3Ge compound by high resolution EXAFS measurements to address the problem of inhomogeneous state, superconductivity and local distortions. The correlated Debye-Waller factor has been taken as an order parameter to determine the temperature dependent local and instantaneous displacements. The choice is due to the fact that the Debye-Waller factor is the parameter which is directly related with the Debye frequency and has direct implication on the superconducting transition temperature. We have studied local lattice displacements as a function of 1) substitutional disorder in the La based superconducting copper oxides and 2) chemical pressure on the CuO_2 plane in the superconducting copper oxides to explore influence on local lattice displacements, the charge inhomogeneous state and the superconductivity. We find a large change in the charge stripe ordering temperature, in addition to the increased static atomic displacements with increasing substitutional disorder. What appears more interesting is the anomalous change of the local Cu-O displacements across the superconducting transition temperature. This change has been found to be dependent on the chemical pressure determined by the micro-strain in the electronically active CuO_2 plane of these complex oxides. While the change at the local displacements at the superconducting transition temperature decreases, the change across the charge stripe ordering temperature gets reduced with increasing micro-strain in the CuO_2 plane. The experiments have direct implication on the correlating electron-lattice interaction, inhomogeneous state and high T_c superconductivity in the complex copper oxides.

We have also measured the local structure of the Nb_3Ge intermetallic compound to explore a possible correlation between the local displacements and the short coherence length superconductivity in this system. In this case, the order parameter of the local displacements is the correlated Debye-Waller factor of the Ge-Nb bonds (σ^2_{Nb}). Temperature dependent local displacements show a small but clear change, appearing as a drop in the σ^2_{Nb} at the superconducting transition temperature T_c, is similar to the one found for the high T_c copper oxides. The comparison of the drop at T_c in the Nb_3Ge intermetallic compound with the one observed in the cuprate superconductors reveals that the local displacements are tied to the superconductivity of the materials with small coherence length. At this stage it is speculative to predict on the precise role of the local lattice displacements, however, it appears that these displacements control the fundamental electronic band structure near the singularity point (M point in the cuprates and Γ point in the A15 intermetallics).

Acknowledgements

This research has been supported by the Istituto Nazionale di Fisica della Materia (INFM), by "progetto cofinanziamento Leghe e composti intermetallici: stabilità termodinamica, proprietà fisiche e reattività" of MURST and by Progetto 5% Superconduttività del Consiglio Nazionale delle Ricerche (CNR).

References

1. *Stripes and Related Phenomena*, eds. A. Bianconi, and N.L. Saini, (Kluwer Academics/ Plenum Publishers, New York, 2000); also *see e.g.* the speciale issues of *J. Superconductivity* vol. **10**, No.4 (1997) and *Int. J. Mod. Phys. B* vol. **14**, No.29-31 (2000).
2. A. Bianconi, G. Bianconi, S. Caprara, D. Di Castro, H Oyanagi, N. L. Saini, *J. Phys. Condens. Matter*, **12** 10655 (2000) and references therein.
3. T. Egami and D. Louca, *Phys. Rev. B* **65**, 094422 (2002) and references therein.
4. S. R. Shenoy, T. Lookman, A. Saxena, and A. R. Bishop, *Phys. Rev. B* **60**, R12537 (1999).
5. A. Lanzara, G.-m. Zhao, N.L. Saini, A. Bianconi, K. Conder, H. Keller and K.A. Müller, *J. Phys.:Condens. Matter* **11**, L541 (1999).
6. R. P. Sharma, S.B. Ogale, Z.H. Zhang, J.R. Liu, W.K. Wu, B. Veal, A. Paulikas, H. Zhang and T. Venkatesan, *Nature* **404**, 736 (2000) and references therein.
7. E. S. Bozin, G. H. Kwei, H. Takagi, and S. J. L. Billinge, *Phys. Rev. Lett.* **84**, 5856 (2000) and references therein.
8. R.J. McQueeney, Y. Petrov, T. Egami, M. Yethiraj, G. Shirane and Y. Endoh, *Phys. Rev. Lett.* **82**, 628 (1999).
9. N. L. Saini, A. Bianconi, and H. Oyanagi, *J. Phys. Soc. Jpn.* **70**, 2092 (2001).
10. A. Bianconi, N.L. Saini, A. Lanzara, M.Missori, T. Rossetti, H. Oyanagi, H. Yamaguchi, K. Oka and T. Ito, *Phys. Rev. Lett.* **76**, 3412 (1996).
11. A. Lanzara, P. V. Bogdanov, X. J. Zhou, S. A. Kellar, D. L. Feng, E. D. Lu, T. Yoshida, H. Eisaki, A. Fujimori, K. Kishio, J.-I. Shimoyama, T. Nodak, S. Uchida, Z. Hussain, and Z.-X. Shen *Nature* **412**, 510 (2001).
12. See e.g. S. R. Shenoy, V. Subrahmanyam, and A. R. Bishop, *Phys. Rev. Lett.* **79**, 4657 (1997) and references therein.
13. *X Ray Absorption: Principle, Applications Techniques of EXAFS, SEXAFS and XANES* edited by R. Prinz and D. Koningsberger, J. Wiley and Sons, New York 1988.

14. A. Bianconi, N.L. Saini, T. Rossetti, A. Lanzara, A. Perali, M. Missori, H. Oyanagi, H. Yamaguchi, and Y. Nishihara, D.H. Ha, *Phys. Rev. B* **54**, 12018 (1996).
15. A. Lanzara, N.L. Saini, A. Bianconi, J.L. Hazemann, Y. Soldo, F.C. Chou and D.C. Johnston, *Phys. Rev.B* **55**, 9120 (1997).
16. N.L. Saini, A. Lanzara, H. Oyanagi, H. Yamaguchi, K. Oka and T. Ito and A. Bianconi, *Phys. Rev. B* **55** 12759 (1997).
17. A. Lanzara, N.L. Saini, M. Brunelli, F. Natali, A. Bianconi, P.G. Radaelli, S-W. Cheong, *Phys. Rev. Lett.* **81**, 878 (1998).
18. N. Ichikawa, S. Uchida, J. M. Tranquada, T. Niemöller, P. M. Gehring, S.-H. Lee, and J. R. Schneider, *Phys. Rev. Lett.* **85** 1738 (2000) and references therin.
19. A. W. Hunt, P.M. Singer, K.R. Thruber, T. Imai, *Phys. Rev. Lett.* **82**, 4300 (1999); P.M. Singer, A. W. Hunt, A.F. Cederström, and T. Imai, *Phys. Rev.B* **60**, 15345 (1999).
20. X.J. Zhou, P. Bogdanov, S.A. Kellar, T. Noda, H. Eisaki, S. Uchida, Z. Hussain, and Z.-X. Shen, *Science* **286**, 268 (1999).
21. J.-S. Zhou and J. B. Goodenough, *Phys. Rev.* B **56**, 6288 (1997).
22. T. Noda, H. Eisaki and S. Uchida, *Science* **286**, 265 (1999).
23. P. C. Hammel, B.W. Statt, R.L. Martin, F. C. Chou, D. C. Johnston and S.-W. Cheong, *Phys. Rev. B* **57**, R712 (1998).
24. S. Tajima, T. Noda, H. Eisaki, and S. Uchida, *Phys. Rev. Lett.* **86**, 500 (2001).
25. I. Watanabe, M. Aoyama, M. Akoshima, T. Kawamata, T. Adachi, and Y. Koike, S. Ohira, W. Higemoto, K. Nagamine, *Phys. Rev. B* **62**, R11985 (2000) and references therein.
26. K. A. Müller, Guo-meng Zhao, K. Conder, and H. Keller *J. Phys. Condens. Matter* **10** L291 (1998).
27. A. Ino, C. Kim, M. Nakamura, T. Yoshida, T. Mizokawa, Z.-X. Shen, A. Fujimori, T. Kakeshita, H. Eisaki, and S. Uchida, *Phys. Rev. B* **62**, 4137 (2000).
28. N.L. Saini, H. Oyanagi A. Lanzara, D. Di Castro, S. Agrestini and A. Bianconi, F. Nakamura and T. Fujita, *Phys. Rev. B* **64**, 132510 (2001).
29. Z. Zhai, P. V. Parimi, J. B. Sokoloff, S. Sridhar, and A. Erb, *Phys. Rev. B* **63**, 092508 (2001).
30. P. Wochner, X. Xiong and S.C. Moss, *J. Supercond.* **10**, 367 (1997) and references therein.
31. *Density Waves in Solids* Frontiers in Physics Vol. **89** ed. G. Grüner, (Addison-Wesley, USA, 1994).
32. A. Bianconi, D. Di Castro, G. Bianconi, A. Pifferi, N. L. Saini, F. C. Chou, D. C. Johnston and M. Colapietro, *Physica C* **341-348** 1719 (2000).
33. P. P. Edwards, G. B. Peakok, J. P. Hodges, A. Asab, and I. Gameson in, *High T_c Superconductivity: Ten years after the Discovery* (Nato ASI, Vol. **343**) ed E. Kaldis, E. Liarokapis, and K. A. Müller, (Dordrecht, Kluwer) (1996) p.135.
34. C. N. R. Rao and A. K. Ganguli *Chem. Soc. Rev.* **24**, 1 (1995); J. B. Goodenough, *Supercond. Science and Technology* **3**, 26 (1990); J. B. Goodenough and A. Marthiram *J. Solid State Chemistry* **88**, 115 (1990).
35. J. Müller, *Rep. Prog. Phys.* **43**, 641 (1980).
36. L. R. Testardi, *Rev. Mod. Phys.***47**, 637 (1975).
37. G.S. Brown, L.R. Testardi, J.H. Wernick, A.B. Hallak and T.H. Geballe, *Sol. Stat. Comm.* **23**, 875 (1977).
38. T. Claeson, J.B. Boyce and T.H. Geballe, *Phys. Rev. B* **25**, 6666 (1982).
39. G.S. Cargill III, R.F. Boehme and W. Weber, *Phys. Rev. Lett.* **50**, 1391 (1983).
40. J. B. Boyce, F. Bridges, T. Claeson, T. H. Geballe, G. W. Hull, N. Kitamura, F. Weiss, *Phys. Rev. B* **37**, 54 (1988).

ANELASTIC MEASUREMENTS OF THE DYNAMICS OF LATTICE, CHARGE AND MAGNETIC INHOMOGENEITIES IN CUPRATES AND MANGANITES

F. CORDERO

CNR, Ist. Acustica "O.M. Corbino", Area della Ricerca di Roma Tor-Vergata, Via del Fosso del Cavaliere, I-00133 Roma, Italy, and INFM
E-mail: cordero@idac.rm.cnr.it

A. PAOLONE, C. CASTELLANO AND R. CANTELLI

Dip. Fisica, Università di Roma "La Sapienza", P.le A. Moro 5, I-00185 Roma, Italy, and INFM
E-mail: annalisa.paolone@roma1.infn.it, carlo.castellano@roma1.infn.it, rosario.cantelli@roma1.infn.it

An overview is presented of the information obtained by anelastic spectroscopy on the low frequency dynamics of various types of intrinsic lattice and electronic inhomogeneities in some cuprate superconductors and manganites. Results are presented on the collective and local pseudodiffusive tilt modes of the O octahedra in multiwell potentials in $La_{2-x}Sr_xCuO_4$; the fast dynamics of the local motion is determined by atomic tunneling and direct coupling with the hole excitations. Two anelastic relaxation processes are identified as due to the motion of the hole stripes. One occurring in the cluster spin glass phase is attributed to the motion between the dopant atoms acting as pinning points, and is completely blocked within the LTT structure. Another process around 80 K is attributed to the depinning of the stripes in a disordered state. In $La_{1-x}Ca_xMnO_3$ a glassy dynamic response below the ferromagnetic transition is attributed to the charge-ordered nanodomains.

1 Introduction

Recently, a great interest has been rising around various types of electronic phase separations and intrinsic inhomogeneities at scales of nanometers in superconducting, magnetic and ferroelectric oxides. Although different physical mechanisms are involved, such nanoscale inhomogeneities seem to be responsible or participate in the phenomena of high-T_c superconductivity (HTS), colossal magnetoresistance (CMR) and extremely high dielectric constants. These inhomogeneities can be directly revealed by several techniques, like various types of neutron and electron diffraction, but their dynamics may be difficult to be measured. In what follows, we present the main results from anelastic spectroscopy on the dynamics of the charge stripes and pseudodiffusive lattice modes in multiwell potentials in the HTS of the $La_{2-x}Sr_xCuO_4$ (LSCO) family and of the nanodomains associated with the ferromagnetic/

charge-ordered phase separation in the CMR manganite $La_{1-x}Ca_xMnO_3$.

2 The anelastic spectroscopy

By anelastic spectroscopy it is intended the measurement of the dynamic elastic compliance[1] $S(\omega,T) = \epsilon_\omega/\sigma_\omega = S' - iS''$, namely the ratio between the response strain ϵ_ω when a periodic stress σ_ω with frequency $\omega/2\pi$ is applied; due to the presence of relaxation processes coupled to stress and producing a retarded anelastic strain ϵ^{an}, ϵ_ω is out of phase by an angle δ with respect to σ_ω, and this causes the negative imaginary part in the compliance. The reciprocal of the compliance is the dynamic elastic modulus $M(\omega,T) = S(\omega,T)^{-1} = M' + iM''$; the tangent of the loss angle δ, or elastic energy loss coefficient, is generally indicated as the reciprocal of the mechanical Q:

$$Q^{-1} = \tan\delta = M''/M' = S''/S' \tag{1}$$

All the experimental results reported here are obtained on ceramic samples cut as bars approximately $40 \times 4 \times 0.6$ mm^3, suspended on thin thermocouple wires and electrostatically excited on one of their first three odd flexural modes, whose frequencies ω_i are in the ratios $1 : 5.4 : 13.3$. Sample and excitation electrode constitute a variable capacity in a high frequency oscillator, whose frequency is modulated by the sample vibration; the latter is measured by an amplitude discriminator. From the resonance frequency of the sample one deduces the real part of the Young's modulus $E' \propto \alpha_i \rho \omega_i^2$ where α_i is a geometrical factor and ρ the density, while from the width of the resonance peak or from the decay time of the free oscillations one deduces the Q^{-1}.

The elastic compliance is the strain susceptibility, and therefore is the elastic analogue of the dielectric and magnetic susceptibilities, with σ and ϵ^{an} instead of electric field and polarization or magnetic field and magnetization. The imaginary part of the dynamic susceptibility S'' is related to the spectral density J_ϵ of the strain fluctuations

$$J_\epsilon(\omega,T) = \int dt\, e^{i\omega t} \langle \epsilon(t)\,\epsilon(0)\rangle \tag{2}$$

through the fluctuation-dissipation theorem,

$$S'' = (\omega V/2k_{\mathrm{B}}T)\, J_\epsilon\,. \tag{3}$$

The measuring frequencies are in the kHz range, and therefore only processes with spectral weight near $\omega = 0$ contribute to the anelastic spectra. These are processes of diffusive or relaxation type, with a correlation function decaying as $\langle \epsilon(t)\, \epsilon(0) \rangle = (\Delta\epsilon)^2 \exp(-|t|/\tau)$, where $\Delta\epsilon$ is the change in anelastic strain due to a jump or transition of the diffusing or relaxing entity with characteristic time τ. In these cases the contribution to the imaginary susceptibility is given by the Debye expression

$$S'' \propto \frac{(\Delta\epsilon)^2}{T} \frac{\omega\tau}{1 + (\omega\tau)^2} . \tag{4}$$

The measurements are made at fixed frequency ω_i as a function of temperature, and therefore the anelastic spectrum consists of peaks from the various relaxation processes α at the temperatures $T_{\alpha,i}$ such that

$$\omega_i \tau_\alpha (T_{\alpha,i}) = 1 ; \tag{5}$$

τ^{-1} generally increases with temperature according to the Arrhenius law,

$$\tau^{-1} = \tau_0^{-1} \exp(-E/k_{\mathrm{B}}T) , \tag{6}$$

and therefore at higher frequency ω_i the conditions for the maxima and the whole spectrum are shifted to higher temperature. By plotting $\ln \omega_i = \ln \tau^{-1}(T_i)$ vs T_i^{-1} one obtains a straight line whose slope is the effective barrier or activation energy E for the process.

3 Motion of the O octahedra in local multiwell potentials in LSCO

The cuprates of the $La_{2-x}Sr_xCuO_4$ family are constituted by planar arrays of CuO_6 octahedra; like in most of the perovskites and perovskite related materials, these octahedra are rather unstable with respect to various tilt modes; causing structural phase transformations. These can be rationalized in terms of mismatch between the equilibrium bond lengths in the LaO blocks and those in the CuO_2 planes, so that the CuO_2 planes are subjected to an increasing compressive stress on cooling, which is relieved by the corrugation of the CuO_2 planes and tilting of the octahedra.[2] Doping relieves the mismatch, as demonstrated by the progressive decrease of the transition temperature $T_t(x)$ from the high-temperature tetragonal (HTT) structure to the low-temperature orthorhombic (LTO) one, and by the concomitant decrease of the average tilt angles in the LTO phase.[3] The local potential felt by the octahedra in the LTO and low-temperature tetragonal (LTT) phases has been calculated;[4,5] it has 8 minima corresponding to the possible LTO and LTT tilts of the pseudo c axis of the octahedron toward the [100] and [110] directions,[4] separated by energy barriers of the order of few hundreds kelvins. A remarkable result is that the lattice remains unstable against such tilt modes also well below the temperature T_t for the HTT/LTO transformation.

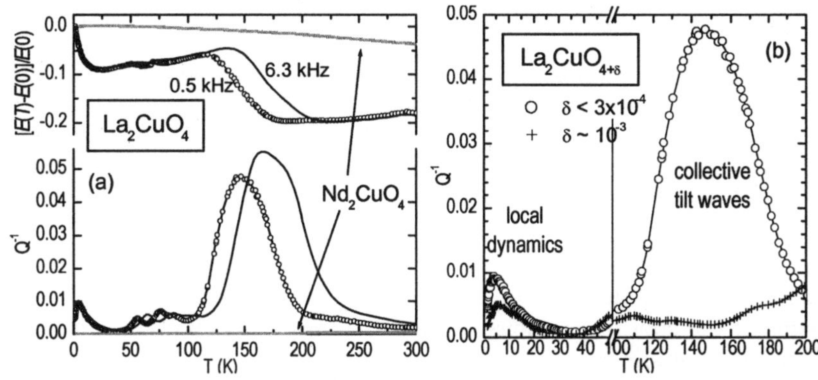

Figure 1. (a) Real and absorption part of the anelastic spectra of $La_2CuO_{4+\delta}$ (measured at two frequencie) and of $Nd_2CuO_{4+\delta}$. (b) Effect of interstitial O on the collective and local pseudodiffusive motions of the O octhedra.

Figure 1a shows the anelastic spectrum of La_2CuO_4 measured at two frequencies, which is dominated by two relaxation processes around 150 K and below 10 K. The intensity of the processes can be better appreciated from the softening in the real part of the modulus $E(T)$ in correspondence with the peaks (18% at 150 K and 10% below 10 K). In fact, concomitant with a peak in S'' and E'' according to Eq. (4), is a step in the real part, $\delta E' \propto -(\Delta\epsilon)^2/T\left[1+(\omega\tau)^2\right]^{-1}$; if there is a broad distribution of τ, the various peaks in the imaginary part are shifted in temperature and produce a broad maximum with height smaller than the sum of the elementary intensities, whereas the steps in the real part are additive. The only possible sources of such large softenings in defect free and stoichiometric La_2CuO_4 are pseudodiffusive modes of the octahedra in the multiwell potential, since the twin walls between the two possible variants of LTO domains contribute to the anelastic spectrum at much higher temperature,[6] and should be immobile below room temperature. For comparison, in Fig. 1a is also shown the completely flat spectrum of Nd_2CuO_4, which has the T' structure without octahedra. Interstitial O constrains several surrounding octahedra to a definite tilt orientation, blocking their hopping to different potential minima; then, by introducing even minute amounts of interstitial O it is possible to clearly distinguish between the collective modes at 150 K, which are completely blocked by $\sim 0.1\%$ of interstitial O, and the local motion below 10 K, which is much less affected. This is shown in 1b. Substitutional Sr disturbs the lattice less than interstitial O, and in fact its effect on the peak at 150 K is much weaker (see Fig. 5).

3.1 *Collective modes: solitonic tilt waves*

The cooperative motion of the octahedra has also been observed as a maximum of the ^{139}La quadrupolar relaxation rate[7] $W_Q(T) \propto 2\tau / [1+(\omega\tau)^2]$ at $\omega/2\pi = 19$ MHz; in fact, $W_Q(T)$ probes the fluctuations of the electric field gradient at the La site, which for undoped La_2CuO_4 are due to the displacements u of the apical atoms of the octahedra, $W_Q(T) \propto \int dt\, e^{i\omega t} \langle u(t)\, u(0) \rangle$. Since u is coupled to the in-plane shear ϵ, both the anelastic and NQR relaxations probe the hopping of the octahedra among the minima of the local potential; at liquid He temperatures the NQR relaxation is dominated by the freezing of the spin fluctuations, so that the faster local motion can only be seen in the anelastic spectra.

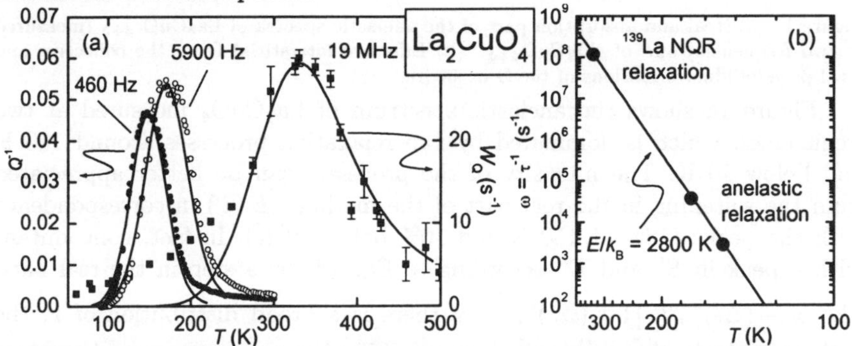

Figure 2. (a) Anelastic and NQR relaxation functions of La_2CuO_4. The curves are fits as explained in the text. (b) Relaxation rates versus the reciprocal peak temperatures.

Figure 2 shows the anelastic and NQR relaxation functions measured on a same La_2CuO_4 sample, and the Arrhenius plot of the mean relaxation rate at the maxima. The activation energy deduced from the slope of $\ln \tau^{-1}$ vs T^{-1} is $E/k_B = 2800$ K, about an order of magnitude larger than the barrier of the local potential calculated for tilting,[4,5] but this is consistent with the cooperative nature of the motion of the octahedra, and can be explained in terms of a one-dimensional model of coupled oscillators,[8] which has solitonic solutions whose spectral density J_u is a central peak, like Eq. (4). The width τ^{-1} of J_u follows the Arrhenius law with an effective barrier for the collective motion which is enhanced over the local barrier by the coupling between the particles.[7] The continuous lines in Fig. 2a are obtained from the above model, with a distribution of τ arising from a gaussian distribution of the values of the coupling parameter. The latter, with a width about 20% of the mean value, corresponds to the distribution in size and shape of the regions where the octahedra clusters build up the cooperative dynamics; these regions are delimited by interstitial O or other defects.

In the original model each particle has two equilibrium positions and the solitons represent propagating domain walls; in the present case there are 8 minima for each octahedron and the solitonic solution may possibly be identified with LTT-like tilt waves that propagate within the LTO domains, as devised by Markiewicz.[9] At lower temperature these tilt waves freeze and their density decreases, as indicated by the decreasing intensity of the anelastic peak in Fig. 2a.

3.2 Local modes: tunneling and coupling to the hole excitations

While doping depresses and hinders the cooperative tilts of the octahedra, due to pinning from the lattice disorder, it has the opposite effect on the fast local motion at liquid He temperatures.[6,10]

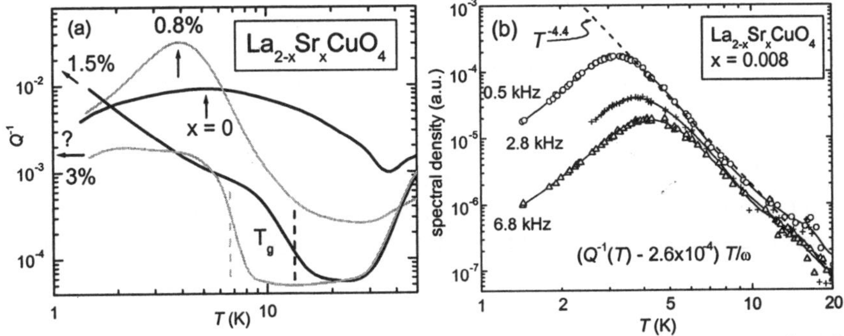

Figure 3. (a) Elastic energy loss of $La_{2-x}Sr_xCuO_4$ at liquid He temperature as a function of doping, measured at ~ 1 kHz. The dashed vertical lines indicate $T_g(x)$ of freezing into the CSG phase for the $x = 1.5\%$ and 3% curves. (b) Spectral density of the local motion of the octahedra deduced from the Q^{-1} curves measured at three frequencies.

The $Q^{-1}(T)$ peak in Fig. 3a shifts from 5.4 to 4 K with only 0.8% Sr doping; its maximum is below 1 K at $x = 0.015$ and not even its tail is visible at $x = 0.03$. The nearly frequency independent steps at T_g are discussed later. The shift to lower temperature and narrowing of the peak is a consequence of a faster relaxation rate $\tau^{-1}(T)$, and of a faster temperature dependence. The process at $x = 0.008$ has been analyzed after subtracting a constant background, and is presented as spectral density of the atomic motion $J \propto T/\omega\, S'' \propto T/\omega Q^{-1}$. The curves measured at different frequencies merge with T^{-n} in the high T side where $\omega\tau \ll 1$ (Fig. 3b), indicating that they can be represented as Eq. (4) integrated over appropriate distributions of $\tau \propto T^{-n}$ and $\Delta\epsilon$. The overall T dependence is $T^{-4.4}$, indicating that the elementary relaxation rates are characterized by a larger exponent, $n \geq 5$.

It appears that a doping of only 0.8% enhances enormously the local lattice fluctuation rate and induces a $\tau^{-1} \propto T^{\sim 5}$ dependence. This fact cannot be understood in terms of overbarrier fluctuations into different potential minima, since a small doping can only change slightly the potential and the phonon spectrum supplying the fluctuation energy. The fluctuations are instead due to atomic tunneling between the minima, with incoherent transitions driven by a direct coupling with the hole excitations.[10] Note that a spectral density of the type of Eq. (4) comes from incoherent transitions between different minima or between different eigenstates delocalized over the minima, but not from the coherent tunneling motion within an eigenstate; the latter still produces a Lorentzian, but centered at the tunneling frequency instead of $\omega = 0$, and has negligible weight at low ω. This type of tunneling dynamics in the presence of charge carriers is well known for H in metals,[11] and leads to $\tau^{-1}(T) \sim T$, or even decreasing with T in case of non-adiabatic coupling to the electrons, and $\tau^{-1} \propto \exp(-\Delta_s/T)$ in the presence of a superconducting gap Δ_s; the $T^{\sim 5}$ dependence indicates that the spectrum of the hole excitations is completely different from that of the free electrons in normal or superconducting metals. This fact is not unexpected, and a more complete study of this relaxation process could supply useful information on these excitations, at least at very low doping.

The anelastic spectroscopy at acoustic frequencies cannot follow the evolution of these fluctuations at higher doping, but doping reduces the mismatch of the bond lengths[2] and therefore makes the minima closer to each other and shallower;[6,10] therefore, one expects that the spectral weight of these fast local fluctuations between minima becomes more and more important, and indeed the peak intensity in Fig. 3a increases of 3.5 times already passing from $x = 0$ to 0.008. It is likely that in the superconducting regime the potential becomes so shallow and anharmonic to give rise to non trivial electron-phonon coupling, as proposed by several authors.[5]

4 The charge stripes in LSCO

In the cuprate superconductors the holes injected into the CuO_2 planes exhibit a peculiar form of phase separation, by segregating into charge rivers or stripes that act as walls between hole-poor domains. This fact has been deduced from the temperature and doping dependence of the correlation length of the antiferromagnetic fluctuations that develop themselves within the hole poor domains, measured by NQR and μSR relaxation.[12,13] At lower temperatures, neutron diffraction experiments reveal arrays of parallel stripes with spacing incommensurate with the lattice constant and determined by doping.[14] The Bragg peaks associated with the charge order can be observed only at doping $x \simeq 1/8$, when the spacing becomes commensurate with the lattice, and with Sr/Nd codoping that stabilizes the LTT structure, whose modulation is parallel to that of the stripes and pins them.[14] Nonetheless, spin correlations as a stripe lattice can be observed over a wide range of Sr doping.[15] Stripes with transverse fluctuations are an ingredient of HTS in some theoretical models,[16] which makes the measurement of their fluctuation dynamics particularly interesting. Such a dynamics must be very fast, but is difficult to be measured; indirect indications come from the observation of the onset of the stripe order into arrays from neutron scattering, μSR, NQR, anelastic and magnetic susceptibility. The onset temperature $T_{\rm on}$ lowers as the the characteristic frequency ω of the probe decreases; this can be intuited by reminding Eqs. (6) and (5) for the condition that the fluctuations become static with respect to the probe. It has also been shown that $T_{\rm on}(\omega, x)$ can be described by means of a broad distribution of activation energies for the fluctuations, with an upper cutoff that decreases with increasing doping.[17]

4.1 Domain walls in the cluster spin glass phase and phase separation at $x < 0.02$

According to the generally accepted phase diagram of LSCO,[18,19] above $x_c \simeq 0.02$ and below $T_g \simeq 0.2/x$ K the Cu^{2+} spins freeze in a cluster spin glass (CSG) phase, where antiferromagnetic correlations are frozen within the hole-poor domains separated by the stripes; the direction of the staggered magnetization is within the *ab* plane but different on different domains, hence the definition of CSG. In this state the stripes are walls between AF domains, and once the spin fluctuations within the domains are frozen, the only response to external fields is the movement of the domain walls. If a magnetoelastic strain $\epsilon^{\rm me}$ is associated with the AF spin ordering, with the principal axes determined by the easy axis of the magnetization, than the CSG phase is an ensemble of domains with different $\epsilon^{\rm me}$. The application of a stress σ, e.g. uniaxial compression along $\hat{\mathbf{n}}$ within the *ab* plane, will enhance the elastic energy $\sigma : \epsilon^{\rm me}$ of the domains with the major principal axis of $\epsilon^{\rm me}$ nearly parallel to $\hat{\mathbf{n}}$ with respect to the other domains; the latter grow at the expenses of the first or equivalently the stripes move in response to a stress.

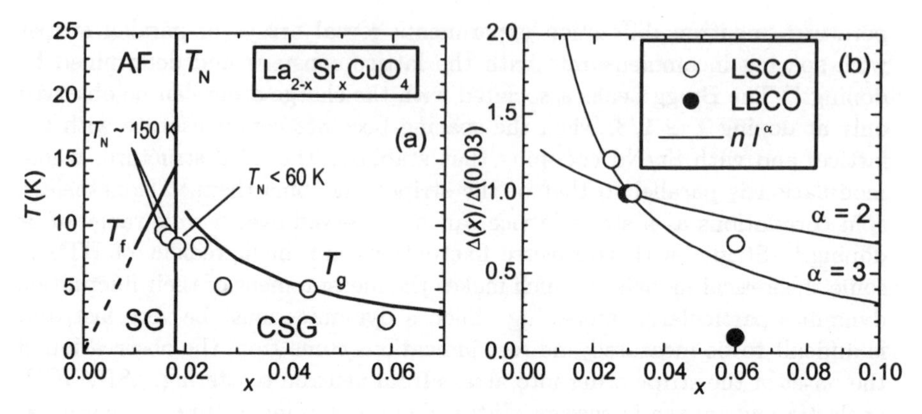

Figure 4. (a) Canonical phase diagram of LSCO from Ref. 19, with the temperatures of the flex of the absorption step attributed to the stripe motion in the CSG phase (circles). (b) Dependence of the amplitude of the absorption step below T_g on doping for $La_{2-x}(Sr/Ba)_xCuO_4$; for the curves see the text.

The $Q^{-1}(T)$ curves of LSCO exhibit a steplike increase below T_g, that has been attributed to the stress induced motion of the stripes within the CSG phase.[20] The step can be seen in Fig. 3a for nominal doping $x = 0.015$ and 0.03, but it has been observed at several dopings up to 0.06. The temperatures of the midpoints of the absorption steps are plotted on the phase diagram of LSCO in Fig. 4a, and correlate very well with $T_g(x)$ (the onset temperatures show an even better agreement but their determination is more arbitrary). Note that they extend well into the region $x < x_c$[21] (recognized by a Néel temperature $\simeq 150$ K), where no spin clusters are believed to exist.[18,19] This observation is in agreement with the recently proposed[22] phase separation for $x < 0.02$ into large domains with $x_1 \simeq 0$ and $x_2 \simeq 0.02$.

Below 10 K the stripes are certainly pinned by various types of defects, starting with the Sr dopants, and indeed it is possible to explain the doping dependence of the intensity Δ of the Q^{-1} increase in terms of models for the anelastic relaxation from dislocations or ferromagnetic (FM) domain walls with density n and mean length l between pinning points,[23] which yield $\Delta \propto nl^\alpha$ with $\alpha = 2$ for dislocations[1] and $n = 3$ for FM walls.[24] The curves in Fig. 4 are obtained from these models assuming that the stripes, with density x, form an array through the Sr atoms equally spaced by $d = 1/\sqrt{x}$ in units of the lattice constant a. The experimental values of $\Delta(x)$ fall between the two cases $\alpha = 2$ and $\alpha = 3$, in agreement with what expected from the motion of DW pinned by the Sr atoms.

The case of doping with Ba, which stabilizes the LTT structure, is different; the anelastic spectrum of the sample with 6% Ba showed a transformation to the LTT structure below ~ 40 K, and a Δ below T_g about 7 times smaller than for 6% Sr. The depression of the absorption step in LBCO with $x = 0.06$ has to be attributed to the blocking of the motion of the stripes parallel to the LTT lattice modulation, even if incommensurate with the lattice (commensuration occurs at $x \simeq 1/8$). The magnitude of the effect indicates that $\sim 87\%$ of the stripes are parallel to the direction of the Cu-O bonds in the LTT phase,[23] in agreement with the direction of the magnetic modulation found by neutron scattering for $x \geq 0.055$.[25].

4.2 Depinning of the disordered charge stripes at high temperature

Above the temperatures where the charge stripes form ordered arrays, a "nematic" phase of disordered stripes should exist.[16] In that temperature range, the processes of depinning from the dopants and/or defects should become possible. Such a process is a typical thermally activated one, with an activation energy determined by the energy barrier for overcoming the pinning center, and should appear as a contribution of the type 4 to the dynamic susceptibility. The anelastic spectrum of LSCO indeed displays some thermally activated processes between 50 K and 100 K, one of which, occurring at ~ 80 K for $\omega/2\pi \sim 1$ kHz, is a good candidate for the stripe depinning.[26,27]

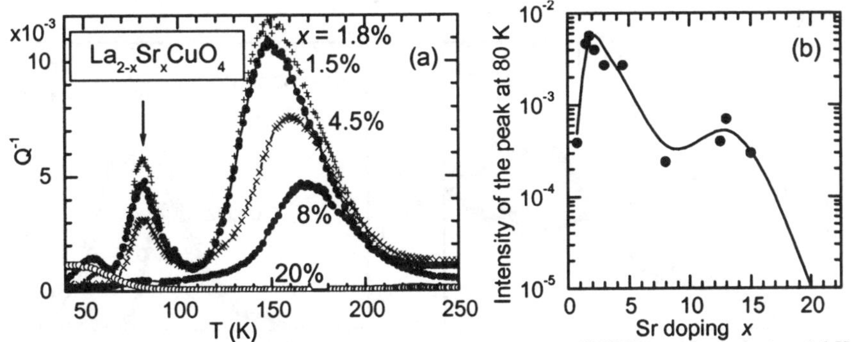

Figure 5. (a) Doping dependence of the anelastic spectrum of LSCO measured at ~ 1 kHz. (b) Intensity of the peak at 80 K as a function of doping; the line is a guide for the eye.

Figure 5a shows the anelastic spectra of LSCO below room temperature for doping up to 20%, with the peak at 150 K due to the tilt waves (see Sect. 3) depressed by the progressive lattice disorder associated with the Sr atoms. The intensity of the peak at 80 K, instead, depends on doping as shown in Fig. 5b.

Both the initial increase with doping and the disappearance in the overdoped state are in agreement with a process of polaronic origin; in addition, a drop of the intensity occurs in the same doping range of the crossover from parallel to diagonal stripes,[25] providing an additional indication that the process is associated with the stripes. The depinning energy deduced from the frequency dependence of the peak is $E_p \sim 0.14$ eV.[27]

5 Glassy dynamics in the phase-separated FM phase of $La_{1-x}Ca_xMnO_3$

The perovskite $La_{1-x}Ca_xMnO_3$ (LCMO) in the range $0.20 < x < 0.50$ undergoes a transition from a paramagnetic insulating (PI) to a metallic ferromagnetic phase (FM) upon cooling below T_C, and exhibits colossal magnetoresistance (CMR), namely a drastic increase of the conductivity on application of a magnetic field. It seems that an important contribution to CMR comes from the coexistence of insulating, probably charge ordered (CO), and metallic domains with sizes of the order of 10 – 50 nm.[28] The application of a magnetic field increases the fraction of the percolating metallic regions, enhancing the CMR effect. Such domains should be rapidly fluctuating, and can be stabilized in manganites with mixed compositions like (La/Pr/Ca)(Mn/Cr)O_3, in which cases features typical of spin glasses are observed.[29] The anelastic spectroscopy provides evidence of glassy dynamics of these domains also in LCMO.

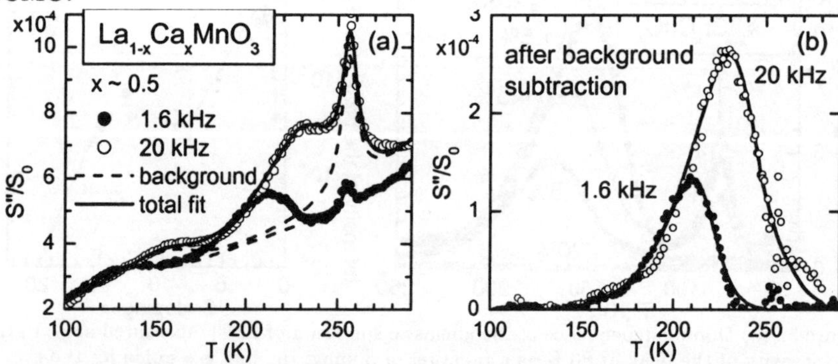

Figure 6. (a) Imaginary part of the elastic susceptibility of $La_{1-x}Ca_xMnO_3$ with $x \simeq 0.5$. (b) After subtraction of the background shown in (a). The continuous lines are fits as described in the text.

Figure 6a presents the anelastic spectrum of $La_{1-x}Ca_xMnO_3$ with doping slightly below the nominal $x = 0.5$ value.[30] The narrow peak near $T_C = 256$ K is due to the PI/FM transition, and is accompanied by a sharp stiffening of the elastic modulus $M' = 1/S'$ (not shown here); the figure reports $S''(T)/S'(0)$ instead of $Q^{-1} = S''(T)/S'(T)$, because in this case the sharp variation of $S'(T)$ makes the $S''(T)$ and $Q^{-1}(T)$ curves inequivalent. The interesting feature is the maximum below T_C, shown in Fig. 6b after the background subtraction; its frequency dependence reminds that of the magnetic and dielectric susceptibilities in spin or proton[31] glasses and relaxor ferroelectrics. In fact, the data can be fitted analogously to the dielectric susceptibility in proton glasses,[31] assuming that the susceptibility S can be expressed in terms of superposition of elementary processes, with a very broad distribution $g(\tau, T)$ of relaxation times obeying the Vogel-Fulcher law $\tau = \tau_0 \exp[E/(T - T_0)]$, so that each contribution like Eq. (4) can be approximated as a δ function, $\delta(\omega\tau - 1)$, and $S''(\omega, T) \simeq \frac{\pi}{2} S(0, T)\, g(\omega^{-1}, T)$. The fit is obtained with $\tau_0 = 5.5 \times 10^{-12}$ s, $T_0 = 193$ K, $E = 620$ K, and demonstrates the glassy nature of the relaxation process near T_C.

The origin of the glassy dynamics should not be sought in broad distributions of the sizes of the CO domains, which seem to have a characteristic dimension of few tens of nm;[28] the spin glass dynamics might rather arise from the competition of interactions with different length scales, which in the manganites include magnetic exchange, Coulomb and elastic interactions.

6 Conclusions

The anelastic spectroscopy can provide information on the low frequency dynamics not only of lattice defects and excitations, but also of extended charge and magnetic structures, through polaronic and magnetoelastic coupling, especially in cases where the other spectroscopies are dominated by other spin or charge excitations.

In LSCO, the charge stripes can be assimilated to walls between AF domains in the CSG phase, and their low temperature fluctuations are limited by the pinning impurities; the pinning centers can be overcome in the disordered state at higher temperature, and the energy barrier is estimated to be ~ 0.14 eV. The coupling between the charge stripes and the lattice likely results in fluctuations of the tilts of the octahedra, but these support both collective and local excitations also in the undoped state; the spectral weight and fluctuation frequency of the local tilts are drastically enhanced by doping, while the collective modes are hindered by the associated lattice disorder.

In the magnetic perovskite $La_{1-x}Ca_xMnO_3$ a signature of glassy dynamics has been found near the transition to the FM state, attributed to the frustrated nature of the competing elastic and electronic interactions responsible for the phase separation into FM and charge ordered domains.

References

1. A.S. Nowick and B.S. Berry, *Anelastic Relaxation in Crystalline Solids.* (Academic Press, New York, 1972).
2. J.-S. Zhou *et al.*, Phys. Rev. B **50**, 4168 (1994).
3. E.S. Bozin *et al.*, Phys. Rev. B **59**, 4445 (1999).
4. W.E. Pickett *et al.*, Phys. Rev. Lett. **67**, 228 (1991).
5. A. Bussmann-Holder *et al.*, Phys. Rev. Lett. **67**, 512 (1991).
6. F. Cordero *et al.*, Phys. Rev. B **57**, 8580 (1998).
7. F. Cordero *et al.*, Phys. Rev. B **59**, 12078 (1999).
8. S. Torre and A. Rigamonti, Phys. Rev. B **36**, 8274 (1987).
9. R.S. Markiewicz, Physica C **210**, 264 (1993).
10. F. Cordero *et al.*, Phys. Rev. B **61**, 9775 (2000).
11. P. Esquinazi, *Tunneling Systems in Amorphous and Crystalline Solids.* (Springer, Berlin, 1998).
12. F. Borsa *et al.*, Phys. Rev. B **52**, 7334 (1995).
13. A. Rigamonti *et al.*, Rep. Prog. Phys. **61**, 1367 (1998).
14. J.M. Tranquada *et al.*, Nature **375**, 561 (1995).
15. S. Wakimoto *et al.*, Phys. Rev. B **63**, 172501 (2001).
16. S.A. Kivelson *et al.*, Nature **393**, 550 (1998).
17. R.S. Markiewicz *et al.*, Phys. Rev. B **64**, 54409 (2001).
18. F.C. Chou *et al.*, Phys. Rev. Lett. **71**, 2323 (1993).
19. D.C. Johnston, *Handbook of Magnetic Materials.* ed. by K.H.J. Buschow, p. 1 (North Holland, 1997).
20. F. Cordero *et al.*, Phys. Rev. B **62**, 5309 (2000).
21. A. Paolone *et al.*, cond-mat/0203051, to be published in Phys. Rev. B.
22. M. Matsuda *et al.*, Phys. Rev. B **65**, 134515 (2002).
23. F. Cordero *et al.*, Phys. Rev. B **64**, 132501 (2001).
24. T. Nattermann *et al.*, Phys. Rev.B **42**, 8577 (1990).
25. S. Wakimoto *et al.*, Phys. Rev. B **61**, 3699 (2000).
26. A. Campana *et al.*, Eur. Phys. J. B **18**, 49 (2000).
27. F. Cordero *et al.*, cond-mat/0206172.
28. M. Fäth *et al.*, Science **285**, 1540 (1999).
29. A. Maignan *et al.*, Phys. Rev. B **60**, 15214 (1999).
30. F. Cordero *et al.*, Phys. Rev. B **65**, 12403 (2002).
31. E. Courtens, Phys. Rev. Lett. **52**, 69 (1984).

NANOSCALE HETEROGENEITY IN THE ELECTRONIC STRUCTURE OF $Bi_2Sr_2CaCu_2O_{8+\delta}$

J.C. DAVIS

LASSP, Dept. of Physics, Cornell University, 531 Clark Hall, Ithaca, NY 14853, USA.

We study the spatially resolved differential tunneling conductance spectra from as–grown and underdoped samples of the cuprate high-T_c superconductor $Bi_2Sr_2CaCu_2O_{8+\delta}$. The data are acquired at 4.2K using atomically-resolved scanning tunneling spectroscopy on the BiO cleave surface of these crystals. For optimally doped samples, the tunneling spectra are consistent with superconductivity over ~99% of the surface but display strong heterogeneity in spectral characteristics with a spatial scale of ~3nm. For underdoped samples, the spectra reveal ~3nm diameter domains with electronic characteristics usually associated with superconductivity that are embedded in an apparently distinct electronic background. If these observations represent the doping dependence of bulk nanoscale heterogeneity in the electronic structure, their trend implies that some form of nanoscale electronic phase separation will occur at very low doping in $Bi_2Sr_2CaCu_2O_{8+\delta}$. Neither the microscopic origin of these phenomena nor the identity of the putative second electronic phase has yet been identified.

1. Introduction

Scanning tunneling microscopy (STM) studies have long provided evidence for nanoscale spatial variations in the electronic characteristics detected at the BiO surface of the cuprate high-T_c superconductor $Bi_2Sr_2CaCu_2O_{8+\delta}$ (Bi-2212)[1,2]. Subsequent to our recent detailed confirmation of these results with high resolution spectroscopic mapping techniques[3], much attention has been directed to this issue[4-8]. The electronic structure detected by STM at the CuO termination surface of $YBa_2Cu_3O_{7-d}$ is also highly heterogeneous at the nanoscale[9]. It is usually assumed that this electronic heterogeneity is purely a characteristic of the CuO plane. However, since the atomic resolution tunneling spectroscopy is not yet available for the superconducting CuO_2 plane of $YBa_2Cu_3O_{7-d}$, it is also possible that its electronic structure displays some nanoscale electronic heterogeneity. Independent of either their universality or microscopic mechanism, nanoscale variations in low energy electronic structure in the cuprates are potentially of great relevance to the phenomenology of high-T_c superconductivity.

Novel theoretical models[10-16] have been developed in response to the recent studies[3-7] of Bi-2212. These models typically analyze strong nanoscale variations in the electrostatic potential energy, the inter-particle interaction energy, or the crystal structure of the cuprates. Such fluctuations could be caused by frustrated electronic phase separation[10,11,12] (as was originally proposed on general principles[17-20]), by unscreened Coulomb effects of dopant-atoms[13,14], by unscreened Coulomb effects of chemical disorder[21], or by self organization of the electronic or crystalline degrees of freedom· Typically in these scenarios, the holes redistribute themselves to minimize the total energy. Thus, at low hole-densities some regions could remain doped while others are relatively undoped. This would create (at least) two types of electronic order, one of which is superconductivity, existing in different nanoscale regions of the same crystal[10-13]. In this paper we report on the status of our STM/STS studies of related phenomena in Bi-2212.

2. Techniques

Scanning tunneling microscopy (STM) can be helpful in the examination of these issues because of its facility to access real-space electronic structure data with atomic resolution. In this work, we use high-resolution scanning tunneling spectroscopy (STS) techniques to compare the superconducting electronic structure on samples[22] of Bi-2212 which are either as-grown or underdoped. The samples are single crystals grown by the floating zone method; the 'as-grown' samples have a mean hole-dopant level $p \approx 0.18 \pm 0.02$ while the underdoped samples are oxygen-depleted yielding $T_c \approx 79K$ and $p \approx 0.14 \pm 0.02$. We note that the relationship between the mean hole-dopant density p for a bulk crystal and the local dopant density in a particular nanoscale region, has not yet yet been determined experimentally. The samples are cleaved in cryogenic ultra-high vacuum and yield atomic resolution topographic images on the BiO surface plane. The tip-sample distance is set by using feedback to stabilize the tip position such that a pre-chosen current (typically I=100 pA) exists at a fixed setup sample-bias voltage (typically V_s=-100meV). The differential tunneling conductance g=dI/dV is then measured versus sample-bias voltage V_b at all locations $\vec{r}$ in a given field-of-view. We make detailed spatial maps of the local density of electronic states (*LDOS*) as a function of energy E by using the identity

$$LDOS(\vec{r}, E = eV_b) \propto \left(\frac{dI}{dV} \right)_{\vec{r}, V_b} \quad (1)$$

3. Experimental results

This type of data[7] can be analyzed to yield images representing the energy gap as a function of location (*gapmaps*) where Δ for each *dI/dV* spectrum is defined as:

$$\Delta \equiv \frac{\Delta_+ - \Delta_-}{2} \qquad (2)$$

and $\Delta_{+(-)}$ is the energy of the first peak above(below) the Fermi level. In Figure 1 we show two such gapmaps, each measured on a 560Å square area. They use an identical grey scale. Fig. 1a is typical of underdoped Bi-2212, while Fig. 1b is typical of the slightly overdoped 'as-grown' samples. The as-grown gapmap is dominated by large interconnected areas of low-Δ (light grey), interspersed with inclusions of high-Δ areas (black). The underdoped sample in Fig. 1a reveals compact, spatially distinct, low-Δ regions (light grey) surrounded by interconnected high-Δ regions (black). For the as-grown sample, the mean gap value is $\overline{\Delta}$=35.6meV with standard deviation σ=7.7meV in a data set of N~16,000 gap values, while for the underdoped sample shown $\overline{\Delta}$=50meV and σ=8.6meV from a similarly sized data set. Obviously reducing the oxygen-doping has shifted $\overline{\Delta}$ to higher values in agreement with the general expectation from many other types of studies.

Even though Fig. 1 has spatial resolution of ≈4Å the tunneling spectra still show strong variations from one pixel to the next. To get a true description, higher spatial resolution data are obviously required. Fig. 2a shows a typical gapmap from a high-resolution experiment acquired with 128x128 pixels on the 147Å square area indicated by the white box in Fig. 1a. Figure 2b is a map of the coherence peak-amplitude, A(Δ), from the same region. We use the same grey scale to represent Δ in Fig. 2a and (inverted) to represent A(Δ) in 2b. This helps point out the strong spatial correlations between these two physically different observables. In these figures one sees compact, almost circular, domains with a low value of Δ, and A(Δ) which climbs rapidly from the domain boundary to reach a maximum at its center. We refer to these regions as *α-domains*. For example, an α-domain is identified inside the white circle on Figs. 2a&2b. Different α-domains have different characteristic values of Δ, and multiple α-domains are sometimes aggregated. Between the α-domains are percolative regions characterized by high Δ and low almost-constant A(Δ). We will refer to these spectroscopically distinct areas as *β-regions*.

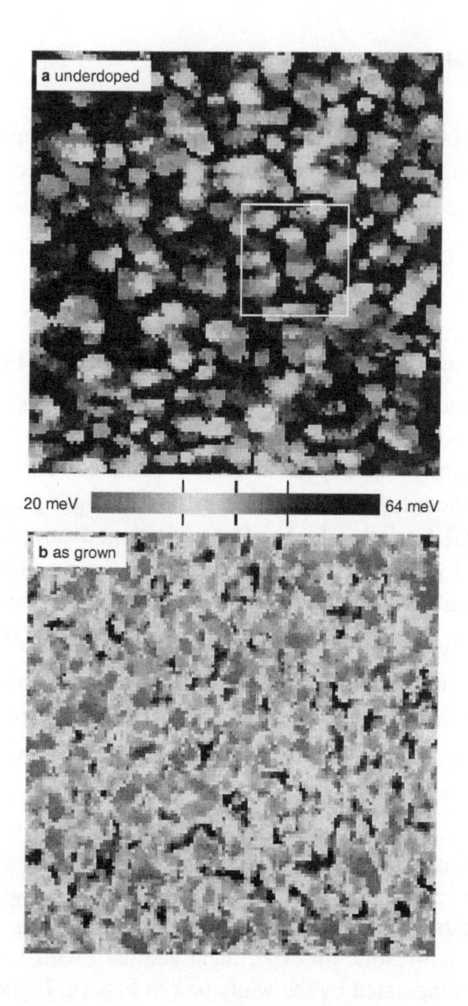

Figure 1 Typical 560 Å square gapmaps from 'underdoped' and 'as-grown' Bi-2212 samples. An identical grey-scale represents Δ in these panels. The grey-scale spans the range 20meV-64meV, and ticks on the scale are placed at 34meV, 42meV, and 50 meV. **a**. Gapmap of an oxygen-underdoped Bi-2212 sample with $T_c \approx 79$ K. This field of view contains approximately 120 compact low-Δ domains which are visible as grey islands against the interconnected black background which are the high-Δ regions. **b**. Gapmap of an as-grown Bi-2212 sample. In this figure the interconnected grey low-Δ regions completely dominate the image. Inclusions of black high–Δ regions appear throughout. Both datasets were acquired with constant current normalization in cryogenic UHV at 4.2K with tunnel junction resistance of 1GΩ set at V = –100mV.

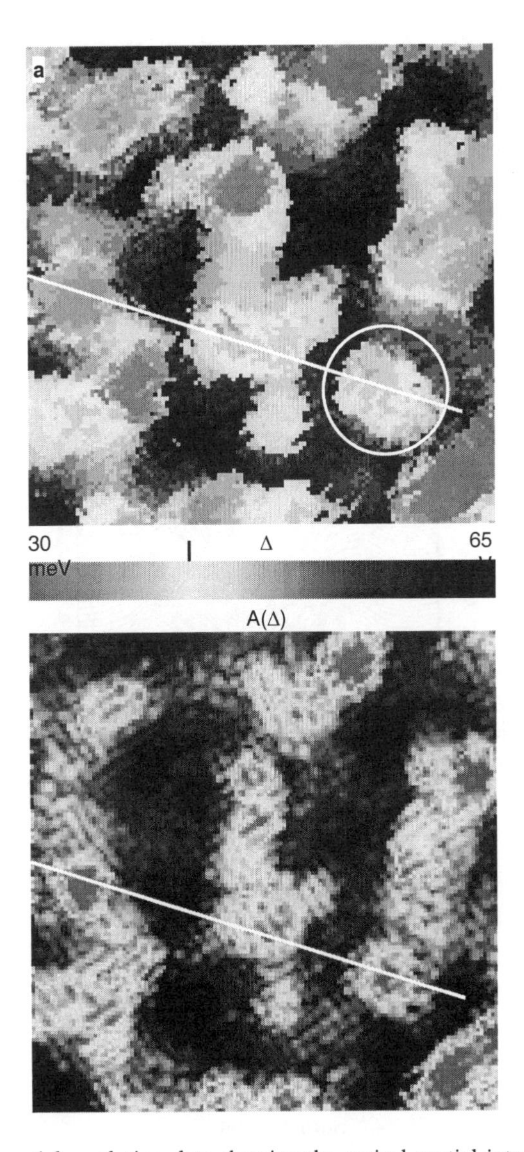

Figure 2 High spatial resolution data showing the typical spatial interrelationship of Δ and A(Δ) for underdoped Bi-2212. The data were measured on the 147 Å square region inside the white box in Fig 1a. **a**. High resolution gapmap revealing approximately 12 α-domains embedded in the percolative β-like background. **b**. Map of A(Δ) at the same location as **a**. Here, to reveal the spatial correlations between Δ and A(Δ), we use the same grey-bar as in **a** but here the scale is inverted and represents A(Δ). Note that A(Δ) is defined as the gap-edge peak at negative bias throughout this paper.

Within each α-domain, the value of Δ varies little, while A(Δ) rises rapidly reaching a maximum in the center. In the surrounding interconnected β-regions, Δ varies slowly while A(Δ) is low and almost constant. To see directly how these phenomena are manifest in the raw dI/dV data, we show in Fig. 3 the unprocessed dI/dV spectra measured along the white line in Figs. 2a&2b. The spectral evolution with passage through three α-domains and the intervening β-regions can be seen.

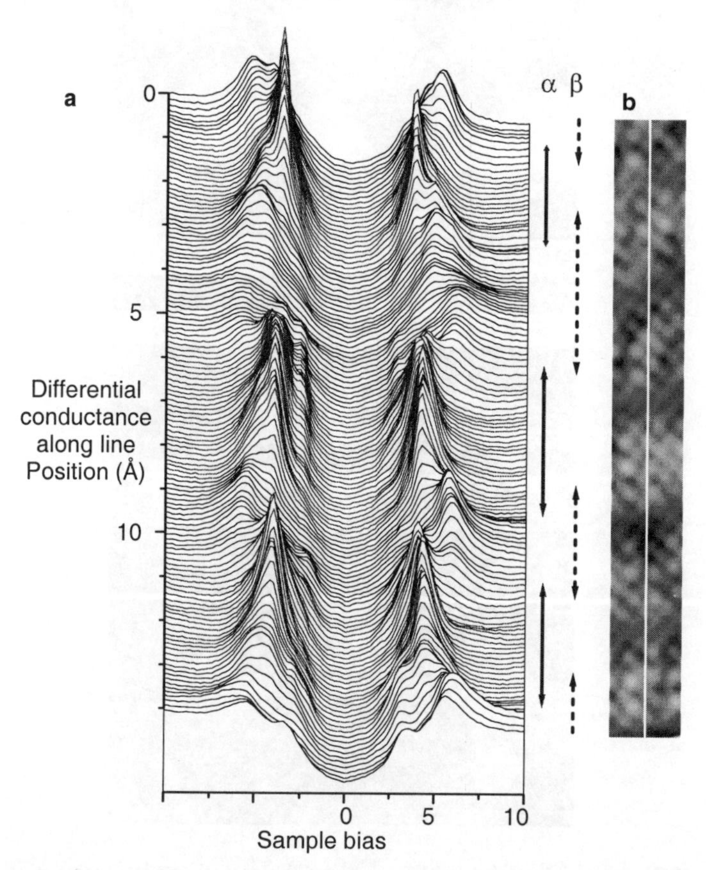

Figure 3 Typical series of dI/dV spectra illustrating how the two distinct types of regions are apparent in the spectra. **a**. These unprocessed spectra were extracted from the same data as Fig.2, along the white line shown in Fig. 2a&b. The spectra are separated by 1.1Å. The α-domain spectra have low gap magnitudes and sharp gap-edge peaks whose amplitude is low at the edges of the domain and rises to a maximum in the center. The β-region spectra have high gap magnitudes and very broad gap-edge peaks whose amplitude is relatively low and constant. **b**. Surface topography with the trajectory along which the spectra in **a,** shown as a white line.

Figures 2&3 demonstrate how the α-domains and β-regions appear spectroscopically distinct. This apparent segregation becomes evident only in significantly underdoped samples because typically $\Delta<50$meV for the α-domains, and underdoped samples have typically $\Delta>50$meV for ~50% or more of their area. On the other hand, as-grown samples have $\Delta>50$meV for less than ~5% and typically ~1%of their area so that the inclusions of β-regions cover a very small fraction of the surface. If these phenomena exist in bulk, the apparent electronic segregation inspires a new picture of the effects of reducing hole-doping in Bi-2212. The α-domains have characteristics usually associated with superconductivity, e.g. sharp gap-edge peaks indicating well-defined Bogoliubov quasiparticles and an increasing areal density associated with rising T_c. In contrast, the β-region spectra exhibit characteristics often associated with the pseudogap phase. Consequently, the emerging picture is reminiscent of a granular superconductor albeit one in material which is not granular in a crystalline sense.

STM is a surface sensitive technique and therefore one may wonder if these phenomena are due only to the condition of the BiO and SrO_2 layers intervening between the CuO_2 layer and the tip. However, there is much evidence for strong nanoscale variations of superconducting electronic structure in studies of in the bulk electronic properties of Bi-2212. This includes: (1) a quasiparticle lifetime at least an order of magnitude shorter[23] in Bi-2212 than in $YBa_2Cu_3O_{7-\delta}$, (2) THz conductivity measurements demonstrating a finite low-temperature σ_1 (ref. 23) that can be explained by strong nanoscale variations in local superfluid density[24], (3) heat capacity measurements showing that with underdoping, $\gamma = C/T = dS/dT$ falls rapidly and the temperature range of the transition widens dramatically[25,26]. (4) NMR studies showing very broad line-widths and additional spin susceptibility at low-temperature indicative of strong electronic disorder[27], (5) inelastic neutron scattering line-widths that are significantly wider in Bi-2212 than in $YBa_2Cu_3O_{7-\delta}$ [28], (6) ARPES studies showing similar evolution of spectra with deliberately introduced bulk disorder as with underdoping[29], (7) interlayer tunneling experiments demonstrating the co-existence of superconducting and non-superconducting energy gaps[30,31], and (8) field dependence of heat capacity revealing characteristic disorder dimension of 6.5nm in Bi-2212[32]. Results from these different techniques would be consistent with each other and with STM data, if nanoscale variations exist in the superconducting properties of *bulk* Bi-2212, and if they are strongly amplified by underdoping.

The STM data discussed here, and the related information from other research programs[4,5,6,8], raise numerous important and, at present, unresolved issues. These include:

1. In Bi-2212, could these phenomena be due to BiO surface preparation/damage or to a spatially disordered tunneling matrix element effect? It seems unlikely because of the similarity of results from many groups using different preparation techniques, because of the strong variations in the characteristic energies of spectroscopic features - an effect which is quite difficult to explain via matrix element disorder, and because of the excellent agreement between Fourier-transform STS and ARPES about the momentum space characteristics of the electronic structure[33]. Nevertheless, these issues need to be resolved.

2. If surface damage is not an issue in Bi-2212, the data (Figs. 1-3) are quite suggestive of granular superconductivity. However, single-particle excitation tunneling spectra cannot definitively distinguish between this scenario and a highly disordered single-phase superconductor. Other techniques are required to demonstrate directly that the amplitude of the superconducting order parameter (or the superfluid density) is heterogeneous on the nanoscale.

3. Can one resolve the apparent contradiction between ~3nm 'granular' electronic structure and the direct demonstration of long correlation-length (>10nm) quantum interference patterns[33] which are concomitant with long coherence lengths for the sub-gap energy quasiparticles? These two observations can be made simultaneously in the same Bi-2212 crystals.

4. Are phenomena like these universal in the cuprates[26]?

5. Are they due to dopant atoms, other chemical impurity effects, or self organizational effects of the crystalline or electronic degrees of freedom?

6. What is the temperature dependence of these phenomena?

7. Is there any net charge density variation associated with these phenomena? This is an important question, but is sometimes overlooked because of confusion about the observables in the STS techniques used. Here the observable is the energy resolved *local density of states*, spatial variations of whose characteristics do not necessarily represent variations in the net charge density within the whole conduction band. New types of experiments, for example

resonant X-ray scattering [34], are required to determine if there is a real charge density heterogeneity.

Much further research into the nanoscale electronic structure of the cuprates will be required before definitive answers can be made to these questions.

Acknowledgments

The author wishes to express his gratitude to all collaborators on this project: H. Eisaki, J. E. Hoffman, E. W. Hudson, K. M. Lang, D.-H. Lee, V. Madhavan, and S. Uchida, This work was supported by the Office of Naval Research under Contract # N00014-00-1-0066, by Grant-in-Aid for Scientific Research on Priority Area (Japan)-a COE Grant from the Ministry of Education and by NEDO.

References.

1 Liu, J.-X., Wan, J.-C., Goldman, A.M., Chang, Y.C. & Jiang, P.Z. *Phys. Rev. Lett.* **67,** 2195-2198 (1991).
2 Chang, A., Rong, Z.Y., Ivanchenko, Y.M, Lu, F. & Wolf, E.L. *Phys. Rev. B* **46,** 5692-5698 (1992).
3 Madhavan, V. *et al.*, *Bull. Amer. Phys. Soc.* **45**, 416 (2000).
4 Cren, T., Roditchev, D., Sacks, W. & Klein, J. *Europhys. Lett.* **54,** 84-90 (2001).
5 Howald, C., Fournier, P. & Kapitulnik, A. *Phys. Rev. B* **64,** 100504/1-4 (2001).
6 Pan, S.H. *et al. Nature* **413,** 282-285 (2001).
7 Lang, K. M. *et al Nature* ***415***, 412 (2002).
8 Matsuda, A. , T Fujii and T Watanabe, to appear in the 'Proceedings of 23rd Int. Conf. Low Temp. Phys.', *Physica B* (2003).
9 Edwards, H.L. *et al.*, *Phys. Rev. Lett.* **73**, 1154 (1994); H.L. Edwards, J.T. Markert, A.L. de Lozanne, *Phys. Rev. Lett.* **69**, 2967 (1992); D. J. Derro, *et al*, *Phys. Rev. Lett.* **88**, 97002 (2002).
10 Ovchinnikov, Y.N., Wolf, S.A. & Kresin, V.Z. *Phys. Rev. B* **63,** 064524 (2001).
11 Ghosal A, Randeria M, Trivedi N., *Phys. Rev. B* 63: 020505 2001.
12 Burgy J, Mayr M, Martin-Mayor V, Moreo A, Dagotto E., *Phys. Rev. Lett.* **87** 277202, (2001).
13 Wang ZQ, Engelbrecht JR, Wang SC, Ding H, Pan SH., *Phys. Rev. B* **65**: 064509 (2002).

[14] Martin, I. & Balatsky, A.V. *Physica C* **357-360**, 46-48 (2001).
[15] Qiang-Hua Wang, Jung Hoon Han, and Dung-Hai Lee *Phys. Rev. B* **65**, 054501 (2002).
[16] Phillips JC, Jung J, *PHILOS MAG* B **81**: 745-756 2001; Jung J, Boyce B, Yan H, et al. *J SUPERCOND* 13: 753 (2000); Phillips JC *PHILOS MAG B* 79: 1477 (1999) .
[17] Gor'kov, L.P. & Sokol, A.V. *JETP* **46,** 420-423 (1987).
[18] Zaanen, J. & Gunnarsson, O. *Phys. Rev. B* **40,** 7391-7394 (1989).
[19] Emery, V.J., Kivelson, S.A. & Lin, H.Q. *Phys. Rev. Lett.* **64,** 475-478 (1990).
[20] Emery, V.J. & Kivelson, S.A. *Physica C* **209,** 597-621 (1993).
[21] H. Eisaki, Private Communication.
[22] We are extremely grateful to Prof S. Uchida of Tokyo University and Dr. H. Eisaki of AIST-Tsukuba for the samples used in these studies.
[23] Corson, J., Orenstein, J., Oh, S., O'Donnell, J. & Eckstein, J.N. *Phys. Rev. Lett.* **85,** 2569 (2000).
[24] Barabash, S., Stroud, D. & Hwang, I.-J. *Phys. Rev. B* **61,** R14924-R14927 (2000).
[25] Loram, J.W., Luo, J.L., Cooper, J.R., Liang, W.Y. & Tallon, J.L. *Physica C* **341-348,** 831-834 (2000).
[26] J. Loram, J. Tallon, W.Y. Liang cond-mat/0212461.
[27] Takigawa, M. & Mitzi, D.B. *Phys. Rev. Lett.* **73,** 1287-1290 (1994).
[28] Fong, H. F. *et al. Nature* **398,** 588-591 (1999).
[29] Vobornik, I. *et al. Phys. Rev. B.* **61,** 11248-11250 (2000).
[30] Krasnov, V.M., Kovalev, A.E., Yurgens, A. & Winkler, D. *Phys. Rev. Lett.* **86,** 2657-2660 (2001).
[31] Suzuki, M. & Watanabe, T. *Phys. Rev. Lett.* **85,** 4787-4790 (2000).
[32] T. Schneider, Private Communication.
[33] Hoffman, J. *et al*, *Science.* 297 1148 (2002); McElroy, K. *et al* to appear in Nature March 2003.
[34] Abbamonte, P. *et al, Science* 297, 581 (2002).

NMR AND μSR INVESTIGATION OF SPIN AND HOLE TEXTURE IN $La_{2-x}Sr_xCuO_4$

PIETRO CARRETTA
Dipartimento di Fisica "A. Volta", Unitá INFM di Pavia,
27100 Pavia, ITALY
E-mail: carretta@fisicavolta.unipv.it

The modifications in the spin dynamic and static properties of $La_{2-x}Sr_xCuO_4$ upon doping are analyzed on the basis of recent NMR and μSR measurements. It is shown how the study of the magnetic properties in the doping range $0 < x \leq 0.12$ yields information also on the charge texture in the CuO_2 planes. In fact, the static and dynamic magnetic properties of $La_{2-x}Sr_xCuO_4$ are closely related to the ones of the antiferromagnetic domain walls generated by the micro-segregation of holes.

1. Introduction

Sixteen years after their discovery [1] several aspects of high T_c superconductors (HT_cSC) remain unexplained. Not only the nature of the microscopic mechanism leading to high temperature superconductivity is still unknown but also the understanding of the normal state excitations is far from being clear. In particular, while several experiments have established that in the lightly overdoped regime there is a quantum critical point (QCP) [2,3], the microscopic nature of the fluctuations leading to the critical behavior, which set in below a characteristic temperature T^*, is still obscure. It is clear that, in analogy to heavy fermion systems, such as as $CeIn_3$ or $CeRhIn_5$ [4], and to β-$Na_{0.33}V_2O_5$ [5], these fluctuations might play a relevant role in the superconducting mechanism [6] (see Fig. 1). On the other hand, it is now well established from a great variety of experiments that HT_cSC show an intrinsic inhomogeneity in the spin, charge and lattice texture [7] resulting from the competition between the superexchange, the elastic and the long-range Coulomb interactions [8]. This inhomogeneous charge distribution may have a well defined topology and the holes may segregate along lines (or stripes) which become static in presence of a strong lattice pinning potential. Although it is well established that static stripes are present in

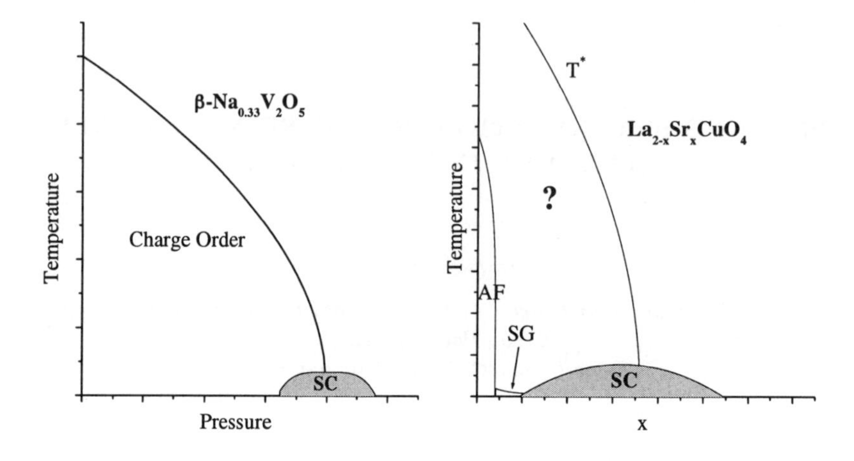

Figure 1. (left) Schematic phase diagram of β-$Na_{0.33}V_2O_5$ after the work by Yamacuchi et al. (Ref. 5) . SC indicates the superconducting phase. (right) Schematic phase diagram of $La_{2-x}Sr_xCuO_4$. The acronyms AF and SG indicate the Néel and spin-glass phases, respectively. The nature of the fluctuations appearing below T^*, which vanishes for $x \simeq 0.19$, is still unclear.

$La_{2-x}Ba_xCuO_4$ for $x = 1/8$ [9], where superconductivity is suppressed, it is not settled if fluctuating charge inhomogeneities, or stripe segments, are present for all doping levels $x > 0$, in particular for the ones leading to high T_c superconductivity, and most of all if these charge fluctuations are the ones diverging at the QCP around $x = 0.19$. A natural approach to investigate these intrinsic miscroscopic inhomogeneities is to use local microscopic probes as nuclei and muons. Therefore, NMR and μSR appear at first glance as ideal tools to investigate the charge inhomogenities and their dynamics in HT_cSC . However, the electric and magnetic hyperfine fluctuations directly driven by the charge inhomogeneities are much weaker than the ones due to the localized spin excitations, so that NMR and μSR are not directly sensitive to the charge fluctuations. Nevertheless, several relevant information on the stripes can be derived by studying their effect on the correlated spin dynamics. In fact, the micro-segregated charges would correspond to the domain walls of antiferromagnetically correlated regions. In the following we will show how, from a series of NMR and μSR experiments in $La_{2-x}Sr_xCuO_4$, from $x = 0$ up to 0.1, information on the static and dynamic properties of the charge distribution can be achieved.

2. The Néel phase: $La_{2-x}Sr_xCuO_4$ for $0 \leq x \leq 0.02$

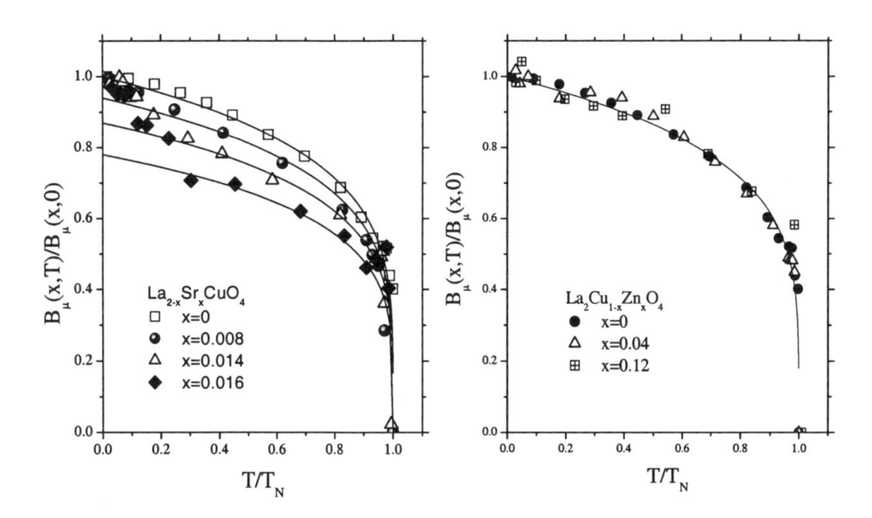

Figure 2. Temperature dependence of the local field at the muon in the Néel phase of $La_{2-x}Sr_xCuO_4$ (left) (see Ref. 10) and $La_2Cu_{1-x}(Mg,Zn)_xO_4$ (right) (see Ref. 12), reported against the reduced temperature T/T_N. The solid lines represent the empirical trend of $B_\mu(x,T) = B_\mu(x,0)(1 - T/T_N(x))^{0.21}$.

An indirect evidence that hole segregation might be present also at low doping levels came from the analysis of the sublattice magnetization in the Néel phase of $La_{2-x}Sr_xCuO_4$, derived from μSR and NQR [10]. In zero field, below the Néel temperature T_N, the local field at the μ^+ (B_μ) is directly proportional to Cu^{2+} magnetic moment [11]. Thus, the temperature (T) dependence of B_μ directly yields the one of the sublattice magnetization. In $La_{2-x}Sr_xCuO_4$, for $x \leq 0.02$ one observes that while for $T \to 0$ $B_\mu(x,T)$ is practically doping independent, a pronounced decrease is present upon increasing T up to $30-50$ K, where the temperature dependence of $B_\mu(x,T)$ recovers the same behavior observed for $x = 0$ (see Fig. 2). One could be tempted to associate this behavior with a motional narrowing effect induced by holes motion, however, if this was the case a peak in the relaxation rate should be observed above $30-50$ K, at variance with the experimental findings. Moreover, above this temperature range, one would expect a difference in the T-dependence of $B_\mu(x,T)$ for different dopings, which was also not observed. On the other hand, the sharp reduction of B_μ on increasing T can be satisfactorly explained by finite size effects on the sublattice magnetization induced by the formation of domain walls [10]. These domains walls can be associated with the hole segregation along stripes. The increase in $B_\mu(x,T)$ below $30-50$ K originates from the hole localization within a single domain wall. In fact, once the holes are localized they form Zhang-Rice singlets and their effect on the sublattice magnetization is expected to be equivalent to the one obtained diluting the CuO_2 planes with $S=0$ impurities, as Zn^{2+} and Mg^{2+} [12,13]. In $La_2Cu_{1-x}(Mg,Zn)_xO_4$ the decrease of $B_\mu(x,0)$ with doping is extremely weak since the antiferromagnetic correlations are able to percolate around the localized singlets and no sizeable effect on the sublattice magnetization curve can be noticed, even up to $x \simeq 0.2$.

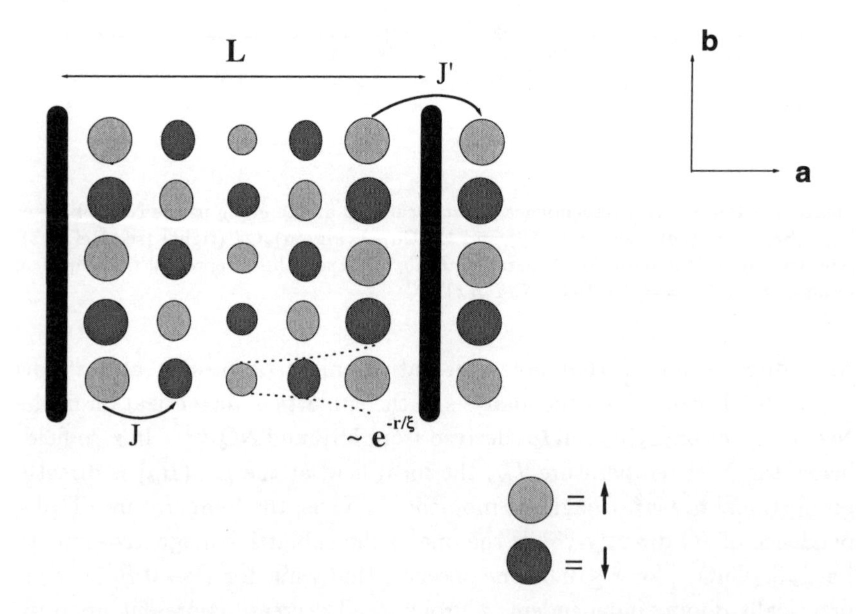

Figure 3. Schematic drawing of the staggered magnetization within an AF cluster. The size of the circles is proportional to the magnitude of Cu^{2+} magnetic moments. J is the supexchange coupling within the cluster while J' is the one between adjacent clusters. ξ is the in-plane correlation length which defines the spatial decay of the staggered magnetization from the domain walls (see text).

3. Coexistence of superconductivity and spin-glass phases: $La_{2-x}Sr_xCuO_4$ for $0.05 \leq x \leq 0.12$

At $x = 0.02$ a first order phase transition separates the Néel phase from the spin-glass phase [14], which in $La_{2-x}Sr_xCuO_4$ extends up to $x \simeq 0.12$, where superconductivity has already set in. Niedermayer et al. [15] have shown by means of μSR that the coexistence of the spin-glass and superconducting phases takes place at a microscopic level. In fact, all the muons injected in the sample were observed to be in the vortex phase when a transverse magnetic field was applied below T_c and also to experience a local field below the spin freezing temperature T_f when a ZFμSR experiment was carried out. The coexistence of both phases at a microscopic level can be understood if holes segregation takes place and yields the percolation of superconductivity around antiferromagnetic (AF) clusters. Hence, the stripes act as domain walls for the AF clusters (see Fig. 3) and their dynamics determines the one of the spins within the cluster.

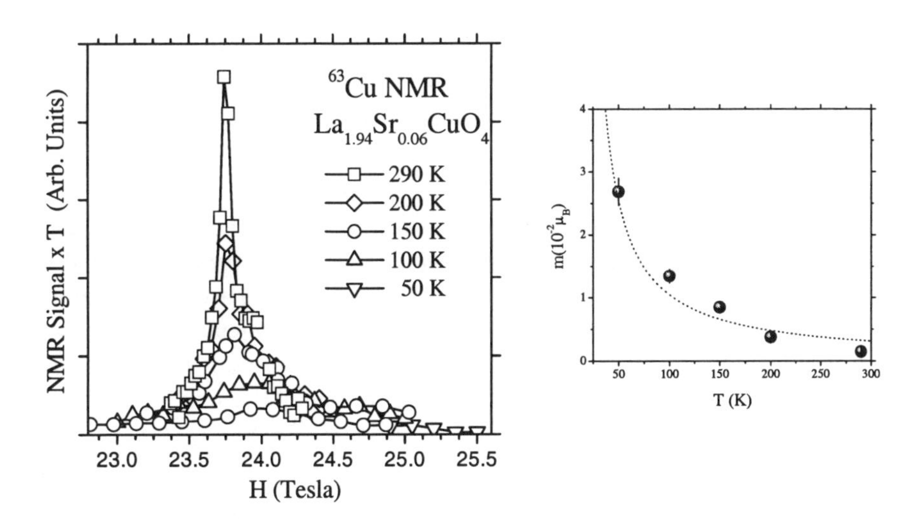

Figure 4. (left) ^{63}Cu NMR spectra in $La_{1.94}Sr_{0.06}CuO_4$ for $\vec{H} \parallel c$ at different temperatures (see Ref. 16). (right) Temperature dependence of the staggered magnetization within an AF cluster derived from the broadening of ^{63}Cu NMR spectra. The solid line shows the Curie-Weiss behavior of the magnetization, with a Curie-Weiss temperature $\Theta = 18 \pm 10$ K.

Information on the microscopic properties of the AF clusters in $La_{2-x}Sr_xCuO_4$, for $0.05 \leq x \leq 0.12$, was achieved by means of ^{63}Cu and ^{139}La NMR. ^{63}Cu NMR spectra show a sizeable broadening on cooling already above 200 K (see Fig. 4) [16]. The broadening cannot be associated with an inhomogeneous distribution of the uniform susceptibility, since for $\vec{H} \parallel c$ the on-site (A) and transferred (B) hyperfine fields would cancel out [17,18]. On the other hand, they will sum up in case of a staggered magnetization and induce a sizeable broadening of the NMR line. This indicates that already at $T \simeq 200$ K there are some AF clusters which are static over the time scale of $\sim 10^{-6}$ s . Therefore, their domain walls must be already partially pinned by the lattice defects at $T \geq 200$ K. One might expect that, as for charge-density waves [19], if these domain walls correspond to stripe segments, the application of a sufficiently strong electric field might cause the sliding of the stripes in the CuO_2 plane. It was observed that for $x = 0.06$ the application of electric fields up to more than 0.1 V/cm did not cause any motional narrowing of the NMR line suggesting a rather strong pinning of the charge inhomogeneities, even around room temperature [20].

From ^{63}Cu NMR linewidth one can derive the T-dependence of the magnetic moments within the cluster. This estimate requires the knowledge of the AF correlation length, which can be derived from ^{63}Cu nuclear spin-lattice relaxation rate $1/T_1$. In fact, by using scaling arguments it is possible to establish a direct relationship between the in-plane correlation length ξ and $1/T_1$ [13]. In Fig. 5 the T-dependence of ξ in $La_{2-x}Sr_xCuO_4$ for $x = 0.06$ is shown. One observes that ξ first increases exponentially on cooling, as expected in the renormalized classical regime [21], and then saturates around room temperature. It must be noticed that ξ saturates at a value $\xi \simeq 1/2x \gg 1/\sqrt{x}$, which directly indicates that the holes are not randomly distributed in the CuO_2 plane but segregate along filaments [22]. This fact evidences that, around 300 K, segments of stripes fluctuate with a characteristic frequency which must be lower than the one of the spin fluctuations within the AF cluster, which is of the order of $Jk_B\sqrt{8S(S+1)/3}/\xi \simeq 10^{13}$ s^{-1}, where J is the superexchange coupling between Cu^{2+} $S = 1/2$ spins. Now, the saturation value for ξ corresponds to the average cluster size (see Fig. 3), which means that all Cu^{2+} magnetic moments within the same cluster have roughly the same average absolute value m. Then m can be directly obtained from ^{63}Cu NMR linewidth ΔH, since $m = \Delta H/2(A+4B)$. Despite of the low accuracy in the linewidth determination one can assert that $m(T)$ follows a Curie-Weiss law with a small Curie-Weiss temperature (see Fig. 4), suggesting a weak magnetic interaction J' among adjacent clusters.

The value of J' can also be estimated on the basis of the analysis of ^{139}La $1/T_1$ below T$\simeq$ 100 K (see Fig. 6). In this temperature range ^{139}La $1/T_1$ diverges exponentially on cooling and reaches a maximum at T_f, where the frequency of the spin fluctuations has reached the MHz range. In analogy to a correlated two-dimensional AF one can write that $1/T_1 \propto exp(2J'(x)/T)$ and derive $J' \simeq 20 - 30$ K, weakly x dependent [23]. Since J' is rather small one might expect, as in standard spin-glasses [24], that the application of a rather high magnetic field, such that the Zeeman energy is larger than the exchange one, would lead to a drastic change both in T_f and in the spin dynamics below 100 K. Remarkably T_f and ^{139}La $1/T_1$ do not show any dependence on the magnet field up to $H = 23$ Tesla [23,25]. This fact is rather surprising and suggests that the coupling among adjacent clusters involves a much larger energy scales as the ones driving the stripes formation or possibly the onset of orbital currents [25].

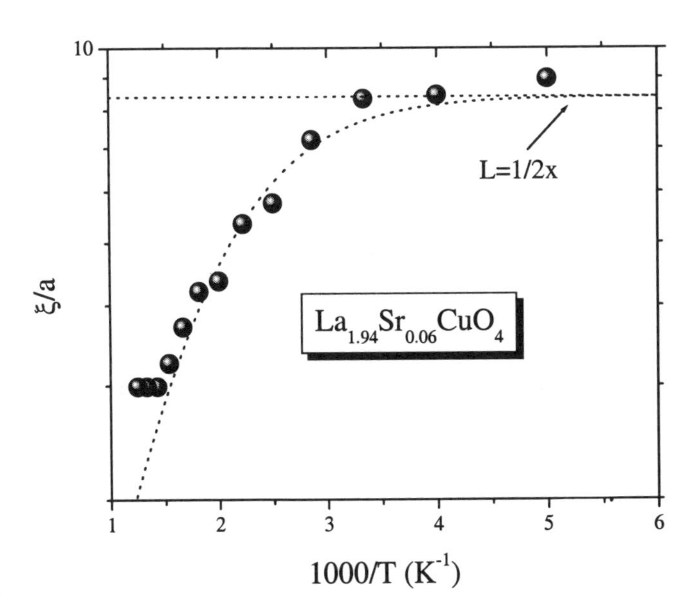

Figure 5. Temperature dependence of the in-plane correlation length derived from ^{63}Cu NQR $1/T_1$ in $La_{1.94}Sr_{0.06}CuO_4$ (see Refs. 16 and 30). $\xi(T)$ follows the phenomenological behaviour $\xi(x,T)^{-1} = L(x)^{-1} + \xi_{RCR}(x,T)^{-1}$, where $\xi_{RCR}(T)$ is the T-dependence expected in the framework of the renormalized classical regime, with a superexchnage coupling $J(x = 0.06) \simeq 1250$ K. The horizontal dotted line shows the value $L(x = 0.06)$ for the average size of the AF clusters.

4. At which temperature charge inhomogeneities appear ?

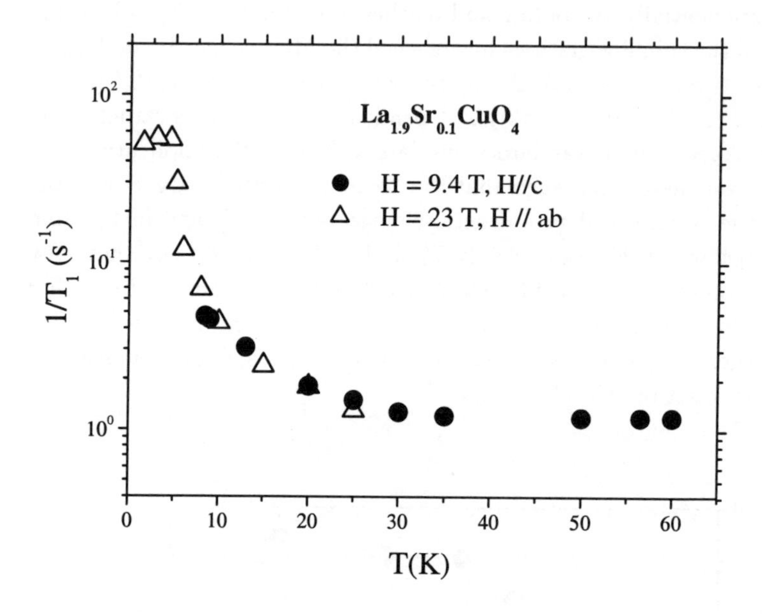

Figure 6. Temperature dependence of ^{139}La $1/T_1$ in $La_{1.9}Sr_{0.1}CuO_4$ for different intensities of the magnetic field (see Refs. 23 and 25). It must be remarked that the low-temperature maximum for $H = 23$ Tesla, which indicates a spin-freezing temperature around 2.5 K, is very close to the one measured by means of ZFμSR (see Ref. 26).

One of the most relevant questions that should be addressed at this point is which is the temperature scale T^* at which charge inhomogeneities appear? Is this temperature scale possibly related to the pseudo-gap opening temperature T_{PG} [27], at which a decrease in the spectral density close to the Fermi energy is observed, as well as a change of curvature in the resistivity and a decrease in the static uniform susceptibility $\chi(0,0)$? Or it corresponds to the spin-gap opening temperature T_{spin} at which a decrease in the low-frequency spectral density at $Q_{AF} = (\pi/a, \pi/a)$ takes place [17,18] ?

From the analysis of the NMR and NQR measurements in $La_{2-x}Sr_xCuO_4$ for $x \simeq 0.06$ one concludes that segments of stripes should be present already around room temperature. In fact, in this temperature range the NMR spectra give evidence for the presence of AF clusters and the correlation length starts to saturate to values around $1/2x$. Thus, it is clear that fluctuating segments of stripes start to form at $T^* \geq 300$ K, well above T_{spin} but in a temperature range possibly close to T_{PG}. Also recent ^{63}Cu NQR $1/T_1$ measurements in $La_{2-x}Sr_xCuO_4$ [29] show an inhomogeneous charge distribution arising well above room temperature. However, it must be stressed that no evidence of a well defined arrangement of the stripes topology along parallel lines has been produced so far, therefore, one is lead to the conclusion that for 200 K T T^* just short segments of stripes are present. These segments fluctuate in the CuO_2 plane with a characteristic frequency below 10^{13} s^{-1} although some of them can be partially pinned by defects. Only at much lower temperatures a well-defined one-dimensional topology for the micro-segregated holes is achieved. This temperature might correspond to the one at which the in-plane resistivity shows an upturn and the correlation among adjacent AF clusters grows, as observed in ^{139}La $1/T_1$. In fact, below this temperature an increase of the in-plane anisotropy in untwinned $La_{2-x}Sr_xCuO_4$ single crystals was observed [28].

5. Summary

It was shown how the microsegregation of holes gives rise to the formation of domain walls for the correlated AF spin dynamics in the CuO_2 planes. This microscopic hole segregation is already present in the lightly doped antiferromagnetic cuprates and extends up to doping levels leading to high T_c superconductivity. It was pointed out how the coupling between adjacent AF clusters, which are separated by hole-rich regions is insensitive to the application of magnetic fields up to 23 Tesla. Moreover, it was shown that dynamical segments of stripes must be present already around room temperature, below the pseudo-gap opening temperature, for $0.05 \leq x \leq 0.12$.

Acknowledgments

The experimental results presented in this manuscript arise from a work done in collaboration with C. Berthier, F. Borsa, A. Campana, M. Horvatić, M.-H. Julien and A. Rigamonti.

References

1. J. G. Bednorz and K. A. Müller, Z. Phys. B 64, 189 (1986).
2. J. L. Tallon and J. W. Loram, Physica C 349, 53 (2001)
3. G. S. Boebinger et al., Phys. Rev. Lett. 77, 5417 (1996)
4. S. Kawasaki et al. Phys. Rev. B 65, 020504 (2002)

5. T. Yamauchi, Y. Ueda and N. Mori, Phys. Rev. Lett. 89, 057002 (2002)
6. C. Castellani, C. Di Castro and M. Grilli, Z. Phys. B 103, 137 (1997)
7. see for example *Stripes and Related Phenomena*, Eds. A. Bianconi and N. L. Saini (Kluwer Academic/Plenum Publisher New York 2000) and references therein
8. A. L. Chernyshev, S. R. White and A. H. Castro-Neto, Phys. Rev. B 65, 214527 (2002)
9. J. M. Tranquada et al., Nature 375, 561 (1995)
10. F. Borsa et al., Phys. Rev. B 52, 7334 (1995)
11. A. Schenck in *Muon Spin Rotation: Principles and Applications in Solid State Physics* (Hilger, Bristol 1986)
12. M. Corti et al., Phys. Rev. B 52, 4226 (1995)
13. P. Carretta, A. Rigamonti and R. Sala, Phys. Rev. B 55, 3734 (1997)
14. M. Matsuda et al., Phys. Rev. B 65, 134515 (2002)
15. Ch. Niedermayer et al., Phys. Rev. Lett. 83, 604 (1998)
16. M.-H. Julien et al., Phys. Rev. Lett. 83, 604 (1999)
17. A. Rigamonti, F. Borsa and P. Carretta, Rep. Prog. Phys. 61, 1367 (1998)
18. C. Berthier, M.-H. Julien, M. Horvatić and Y. Berthier, J. Phys. I 12, 2205 (1996)
19. P. Ségransan et al., Phys. Rev. Lett. 56, 1854 (1986)
20. Yu. A. Dimashko, C. Morais Smith, N. Hasselmann, and A. O. Caldeira, Phys. Rev. B 60, 88 (1999)
21. S. Chakravarty, B. I. Halperin and D. R. Nelson, Phys. Rev. B 39, 2344 (1989)
22. P. Carretta et al., Eur. Phys. J. B 10, 233 (1999)
23. M.-H. Julien et al., Phys. Rev. B 63, 144508 (2001)
24. see for example S. M. Dubiel, K. H. Fischer, Ch. Sauer, and W. Zinn, Phys. Rev. B 36, 360-366 (1987)
25. M. Eremin and A. Rigamonti, Phys. Rev. Lett. 88, 037002 (2002)
26. A. Weidinger et al., Phys. Rev. Lett. 62, 102 (1989)
27. T. Timusk and B. Statt, Rep. Prog. Phys. 62, 61 (1999)
28. Y. Ando et al., Phys. Rev. Lett. 88, 137005 (2002)
29. P. M. Singer, A. W. Hunt and T. Imai, Phys. Rev. Lett. 88, 047602 (2002)
30. S. Fujiyama et al., J. Phys. Soc. Jpn. 66, 2864 (1997)

PHONON MECHANISM OF HIGH-TEMPERATURE SUPERCONDUCTIVITY

T. EGAMI

Laboratory for Research on the Structure of Matter and Department of Materials Science and Engineering, University of Pennsylvania, Philadelphia, PA 19104, USA

E-mail: egami@seas.upenn.edu

The majority opinion on the origin of high-temperature superconductivity is that it is a purely electronic phenomenon involving spins, and phonons are irrelevant to the mechanism. However, such a view is based upon the knowledge of the phonon mechanism in simple metals, while the cuprates are much more complex due to various competing forces. We suggest that the electron-phonon coupling in the cuprates is significantly different because of their covalency, ionicity and strong electron correlation. In particular phonon-induced intersite charge transfer creates a strong electron-phonon coupling that is dependent on mode and wavevector, and could form a basis for phonon mechanism of high-temperature superconductivity.

1 Introduction

The cuprate compounds that exhibit high-temperature superconductivity (HTSC) are the most pronounced example of complex oxides in which multitude of degrees of freedom are competing to create various unusual phenomena. As demonstrated in another example of complex oxides, manganites that show colossal magnetoresistivity (CMR), the coupling between charge and lattice is one of the main forces that define the characteristics of this class of materials. Therefore it is quite perplexing that, in the majority view on the mechanism of HTSC, only spins are important and charge and lattice degrees of freedom are irrelevant. Several observations are usually cited in eliminating phonons from consideration; the small isotope effect, absence of evidence for strong electron-phonon (*e-p*) coupling in the normal state transport properties, the *d*-symmetry, competition against the magnetic mechanism, *etc*. These arguments are, however, largely based upon the knowledge of the *e-p* coupling in simple metals. The cuprates, on the other hand, are far from simple, and the *e-p* coupling can be different in significant ways from that in simple metals, because of their covalency, ionicity and strong electron correlation. In this paper, based upon the results of recent inelastic neutron scattering measurements and model calculations, we suggest that the *e-p* coupling in the cuprates is characterized by strong interband coupling that induces charge transfer and depends upon the phonon mode and momentum. In our model calculation a phonon mechanism based upon such a coupling can create *d*-wave superconductivity that is consistent with the experimental observations and the superconductive transition temperature well over 100 K.

2 Phonons in the Cuprates

2.1 Anomalous phonon modes

Phonons in the HTSC cuprates have been studied by various techniques, such as IR [1], Raman [2] spectroscopy and inelastic neutron scattering [3]. The strength of the *e-p* coupling can be estimated from the renormalization of the phonon frequency near T_C, and the dependence of the phonon dispersion on doping. The picture that emerges from these studies is that the *e-p* coupling in the cuprates is not weak, but not overly strong for most phonon branches. When examining the doping dependence of each mode more carefully, however, strong doping dependence is observed for two modes; the in-plane oxygen half-breathing mode and the apical oxygen *c*-axis mode.

The high-frequency optical phonon modes of the CuO_2 plane polarized in plane are relatively simple Cu-O bond-stretching modes, with displacements mostly by oxygen ions. The zone-center mode is ferroelectric, while the zone-edge modes are the full-breathing mode at $\boldsymbol{Q}$ = (0.5, 0.5, 0) in the units of reciprocal lattice vectors, and the half-breathing mode at $\boldsymbol{Q}$ = (0.5, 0, 0). In the undoped cuprates, such as La_2CuO_4 or $YBa_2Cu_3O_6$, both the LO and TO modes are nearly dispersionless. But upon doping the half-breathing mode strongly softens [3]. Our recent pulsed neutron inelastic scattering measurement of this mode (Fig. 1) for $YBa_2Cu_3O_{6.95}$ carried out with the MAPS spectrometer at the ISIS [4,5] shows that the softening is strongly anisotropic, 12 % along ***b*** (parallel to the Cu-O chain), and 24 % along ***a*** (perpendicular to the chain). On the other hand, the TO mode remains dispersionless even with doping.

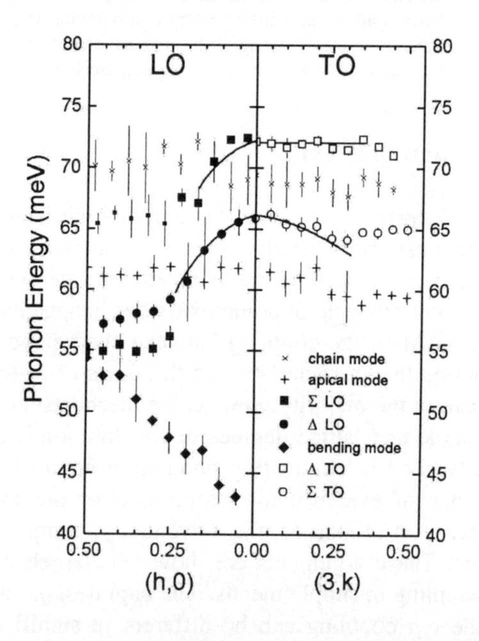

Fig. 1. Dispersion of in-plane Cu-O bond-stretching modes for $YBa_2Cu_3O_{6.95}$ at T = 110 K determined with MAPS at the ISIS [4,5]. The Σ mode has polarization along ***a*** (perpendicular to the chain), while the Δ mode is polarized along ***b*** (parallel to the chain). The Σ LO mode softens by 17 meV towards the zone boundary, while the Δ LO mode softens only by 8 meV.

In YBCO there are two *c*-axis apical oxygen modes, the IR mode in which two apical oxygen ions above and below the double CuO_2 layers move in phase, and the Raman mode in which they move out of phase to each other. Both of them soften by about 10 % by doping up to the optimum doping, but they remain nearly dispersionless.

2.2 *Temperature dependence*

The in-plane Cu-O bond-stretching LO mode in $YBa_2Cu_3O_{6.95}$ was found to show anomalous temperature dependence. An order parameter, defined as the average scattering intensity at $\boldsymbol{Q}$ = (3.25, 0, 0) between 51 and 55 meV minus that between 56 and 68 meV, is shown in Fig. 2. It has a striking resemblance to the superconducting order parameter, with a sharp change around T_C = 93 K, suggesting a direct link between this mode and superconductivity.

The energy of the apical oxygen Raman mode (62 meV in $YBa_2Cu_3O_{6.95}$) is in the same energy range as the in-plane LO mode. Since they belong to the same irreducible representation, they mix with each other. When they mix in-phase they represent an oxygen breathing mode around Cu (Fig. 3 (a)), while if they are out-of-phase they are a Jahn-Teller (JT) mode (Fig. 3 (b)). Note that the distance between the apical oxygen ion and the in-plane oxygen ions become bifurcated for the breathing mode, while it remains little affected for the Jahn-Teller mode. Some time ago we observed with the pulsed neutron atomic pair-density function (PDF) analysis of $Tl_2Ba_2CaCu_2O_8$ that the O-O distance of 3.2 Å becomes split into 3.0 amd 3.4 Å near T_C [6]. The PDF anomaly is dynamic, since it depends upon the detector angle. This observation now can be explained by softening of the breathing mode around T_C.

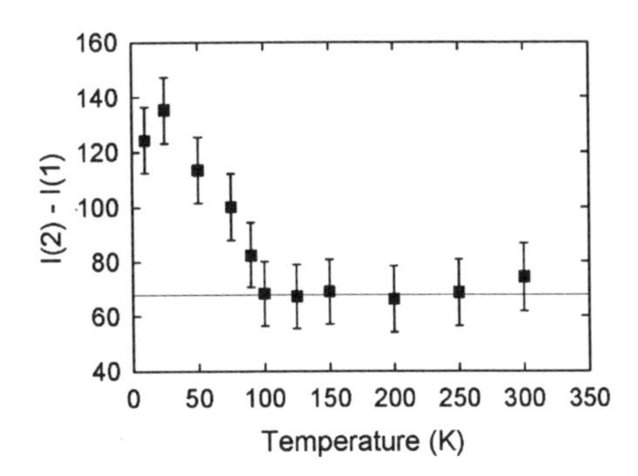

Fig. 2 $\Delta I = I(q = 0.25, \omega = 51 - 51 \text{ meV}) - I(q = 0.25, \omega = 56 - 68 \text{ meV})$ as a function of temperature for $YBa_2Cu_3O_{6.95}$ [4,5]. T_C is 93 K.

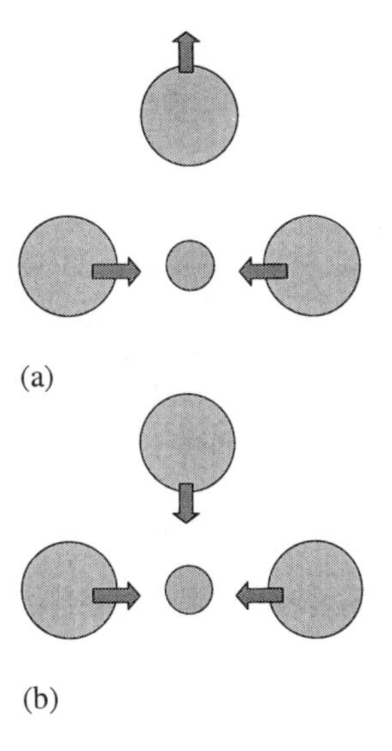

Fig. 3 Mode of coupling for the in-plane and c-axis modes: (a) JT-mode, (b) Breathing-mode. Large circles represent oxygen and small circles copper.

2.3 Extended Lyddane-Sachs-Teller relationship

Our attempts to model the observed phonon dispersion with the lattice dynamics (shell model) convinced us that it is difficult to explain the observed strong softening of the Cu-O bond-stretching LO mode at the zone boundary and the absence of such softening in the TO mode within the framework of simple lattice dynamics. To explain the softening a rather unphysical attractive potential (a spring with a negative spring constant) has to be present between O ions, either across Cu or between neighboring O in the plane. Furthermore, such an interaction must be present ONLY for the LO mode, but not for the TO mode. This awkward lattice dynamics suggests an electric or electronic origin of such an interaction.

A well-known example of such an electric field effect for optical phonons is the Lyddane-Sachs-Teller (LST) relationship between the LO and TO frequencies at $\boldsymbol{q}$ = 0 [7];

$$\left(\frac{\omega_{LO}}{\omega_{TO}}\right)^2 = \frac{\varepsilon_0}{\varepsilon_\infty} \tag{1}$$

where ε_0 and ε_∞ are the low and high frequency dielectric constants. This relation could be extended to finite $\boldsymbol{q}$,

$$\omega_{LO}^2(\boldsymbol{q}) = \omega_{TO}^2(\boldsymbol{q}) + \frac{\omega_{LO}^2 - \omega_{TO}^2}{\varepsilon(q,\omega)} \tag{2}$$

where ω_{LO} and ω_{TO} are the bare LO and TO frequencies without the *e-p* coupling, and $\varepsilon(\boldsymbol{q}, \omega)$ is the electronic dielectric function [8]. If we apply this relationship the fact that the LO frequency is lower than the TO frequency translates into a negative $\varepsilon(\boldsymbol{q}, \omega)$. A negative dielectric constant occurs in several circumstances, such as over-screening associated with anti-resonance. For regular screening eq. (2) yields smaller renormalization due to the *e-p* coupling, since $\varepsilon > 1$. However, when $\varepsilon < 0$, $|\varepsilon|$ can be smaller than unity, and the phonon renormalization can be very large, resulting in a strong *e-p* coupling and a possibility of high values of T_C, as suggested by Ginzburg some time ago [9]. Tachiki and Takahashi [8] assumed such anti-resonance and a negative dielectric function in their vibronic mechanism of superconductivity.

3 Electron-Phonon Coupling in the Cuprates

3.1 Electronic polarization in ferroelectric oxides

An alternative, more microscopic description of the LO mode softening is to consider phonon-induced charge transfer. In order to explain this mechanism let us make a detour to the case of ferroelectric oxides. It has been known for some time that strong ferroelectricity in transition metal oxides, such as $BaTiO_3$, is not only due to ionic polarization but also due to strong electronic polarization as well [10-12]. For instance, in titanates Ti^{4+} has a nominal d^0 configuration, but strong covalent Ti-O bonding hybridizes the Ti-*d* and O-*p* orbitals. Now, let us consider the case where Ti ions move to the right and O to the left (Fig. 4). This motion creates a positive ionic polarization pointing to the right. But since the distance between Ti and O is reduced the hybridization between Ti-*d* and O-*p* is increased (covalency of the Ti-O bond is strengthened). As a result electrons are partially transferred from the O-*p* orbital to the Ti-*d* orbital (Fig. 4). Since this charge transfer reduces ionic charges of both Ti and O, it may appear that the electronic polarization is weakened. However, Resta showed that the polarization is correctly described as the total current due to the displacement [13]. The electron transfer from O to Ti in the left means a current to the right, thus polarization towards the right. In this argument the electronic polarization adds to the ionic polarization and enhances the total polarization. The total polarization is,

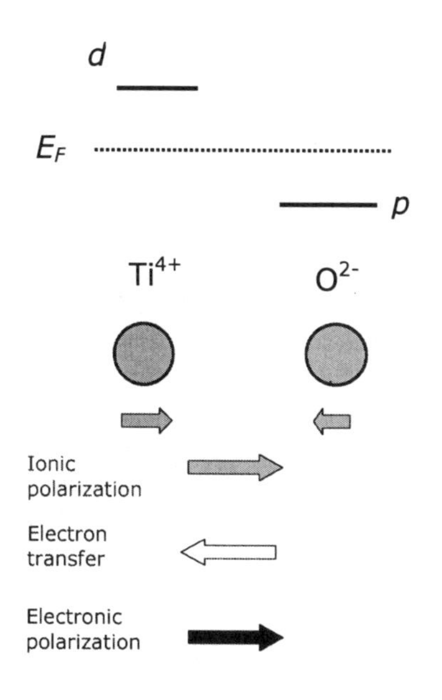

Fig. 4 Phonon-induced charge transfer in Ti-O. As the Ti-O distance is reduced electron is transferred from O to Ti, creating a positive current from Ti to O, which represents the electronic polarization.

$$P = Zu + \Delta Za = \left(Z + \Delta Z \frac{a}{u} \right) u = Z^* u \qquad (3)$$

where u is the ionic displacement, a is the interatomic distance and Z^* is Born effective charge. Since a/u can be as large as 10 – 20, Born effective charge can be twice as large as the nominal charge. The electronic polarization was further identified with the Berry phase of the system associated with ionic displacements by King-Smith and Vanderbilt, who developed an elegant expression of the polarization in terms of single electron wavefunctions [14].

It is worthwile to note that the electronic transitions induced by the optical phonons are interband transitions, $\boldsymbol{k} \rightarrow \boldsymbol{k}' + \boldsymbol{G}$, where $\boldsymbol{G}$ is a reciprocal lattice vector. Since the band gap is considerable (3 – 5 eV), all the $\boldsymbol{k}$ states contribute more or less equally. It is important, in reference to the *e-p* coupling in the cuprates, that strong electron-lattice coupling is produced in the insulating ferroelectrics, even without Fermi surface nesting, by the interband transition.

3.2 Electronic polarization in the cuprates

In the undoped cuprates, such as La_2CuO_4 or $YBa_2Cu_3O_6$, the Cu-O bond-stretching phonons induce electron transfer from the O-*p* state to the Cu upper Hubbard *d*-level, resulting in a current from Cu to O, just as the titanates. However, when the cuprates are doped, doped holes occupy mainly the O-*p* levels. When the Cu-O distance is reduced holes, rather than electrons, flow from O to Cu, either to the lower Hubbard d_{x2-y2} band or the d_{z2} band, thus the direction of the current is from O to Cu, and the electronic polarization is anti-parallel to the ionic polarization, reducing the total polarization.

The electronic polarization was calculated for the 1-dimensional Hubbard model (6 Cu + 6 O, or 4 Cu + 4 O) with the cyclic boundary condition and the Su-Schrieffer-Heeger *e-p* coupling [15],

$$H = \sum_{i,j,\sigma}\left(t_{ij} + \frac{\alpha}{a}\left(u_i - u_j\right)\right)\left(c^+_{i,\sigma}c_{j,\sigma} + c^+_{j,\sigma}c_{i,\sigma}\right) + \sum_i U n_{i\uparrow} n_{i\downarrow} \tag{4}$$

where u_i is the displacement of the *i*-th ion [16]. As shown in Fig. 5 the effective charge of oxygen is strongly dependent on $\boldsymbol{q}$, the phonon wavevector. It increases negatively for $q = 0$ with α, the strength of the *e-p* coupling, but decreases for $q = \pi$ and changes its sign if α is strong enough to be appropriate for the cuprates. Such inversion was observed also for the model system describing titanates, if the Ti upper Hubbard level and the O level crossover [17,18]. Thus at $q = \pi$ the electronic polarization is anti-parallel to the ionic polarization. This means that the electronic dielectric constant $\varepsilon(\boldsymbol{q}, \omega)$ is negative at $q = \pi$, and justifies the scenario by Tachiki and Takahashi [8]. It is important to note that the strength of this coupling originates from the charge transfer due to interband transition, $\boldsymbol{k} \rightarrow \boldsymbol{k'} + \boldsymbol{G}$. The amplification of the *e-p* coupling by multi-band effect was also

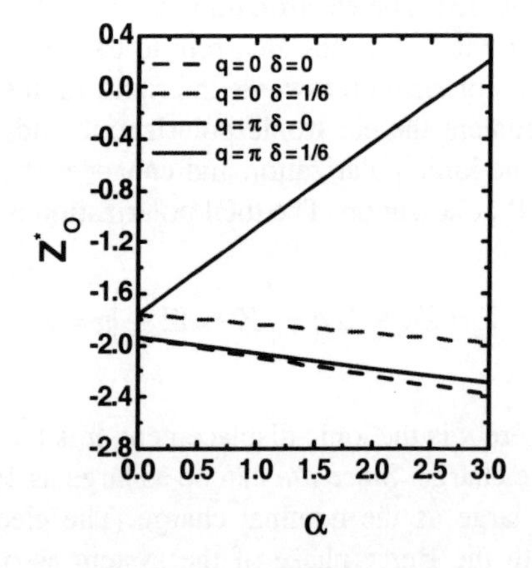

Fig. 5 Born effective charge of oxygen at $q = 0$ and π (0.5) as a function of the e-p coupling parameter. $\alpha = 3$ is appropriate for the cuprates [16].

pointed out by Weger [19]. The possible involvement of charge transfer excitation in HTSC has been suggested by Little [20,21].

3.3 Phonon mediated superconductivity

Motivated by these results the Eliashberg equation was set up and solved in the vicinity of T_C, by assuming a model band structure determined to fit the results of the photoemission measurement and the dielectric function determined from the phonon dispersion data shown in Fig. 1 using the equation (2). The electronic dielectric function was assumed to be negative only in the parts of the q-space around $(\pi, 0)$. The symmetry of the regions with negative dielectric function, where the e-p coupling is strong, is consistent with the d-wave superconductivity. Values of T_C as high as 200 K were obtained for d-wave superconductivity [22]. Since the results are strongly dependent on details of the band structure, the values of T_C should be taken with elements of caution. But it is important that such a high value of T_C was easily obtained with the Eliashberg equation. The anisotropic s-wave solution is also close in energy, and depending on the details of the parameters the ground state can be either d-wave or s-wave. In the case of the d-wave solution this mechanism does not compete against the magnetic mechanism. In the presence of strong spin fluctuations the magnetic mechanism and the phonon mechanism can co-exist, with the relative strength depending on the hole concentration.

4 Discussion

The difficulty of achieving high values of T_C with the phonon mechanism has been discussed by many. It has been generally believed, based on the McMillan equation [23], that the maximum value of T_C with the BCS theory is around 30 K. The most critical problem has been that the strong e-p coupling results in covalent bonding and structural phase transformation [24]. In the present mechanism, however, the phonon-induced charge transfer, transfer of holes from oxygen to copper, actually *weakens* covalency and increases ionicity. This prevents the system from the ferroelectric catastrophe of structural transformation, since this softens the LO phonons rather than the TO phonons as in the case of ferroelectric oxides, and the LO phonons are more difficult to soften completely than the TO phonons because of the internal electric field it creates. At the same time it results in a negative dielectric function, and a value of $1/|\varepsilon| > 1$ that produces amplification of the local potential.

Another problem of the phonon mechanism is formation of polarons that increases an effective mass of an electron and reduces the Bose condensation temperature [25]. For the cuprates the equivalent of polaron formation is the formation of spin-charge stripes. Indeed the system has a tendency for stripe formation that competes against superconductivity [26]. The stability of polarons and stripes depends upon the ionic size [27], and they can be suppressed by choosing the elements with the appropriate ionic radii. Thus we believe that the phonon-induced charge transfer mechanism can results in values of T_C much higher than the conventional limit.

5 Conclusions

The phonon mechanism was abandoned very early, almost right after the discovery of HTSC, in the frantic search of a novel and exotic mechanism of HTSC. However, while none of the exotic mechanisms of HTSC have been proven right, evidences that phonons are closely involved in the mechanism of HTSC are accumulating. In this paper we suggested that the arguments against the phonon mechanism are based upon the knowledge of the electron-phonon coupling in simple metals, and cannot be applied to the cuprates. In the cuprates some phonons induce intense charge transfer between ions, and couple strongly to charge. A model calculation based upon this scenario demonstrated that the superconductive transition temperature over 100 K can be achieved within a reasonable range of parameters. In our view phonons should be taken more seriously as a possible agent of HTSC.

6 Acknowledgements

The author thanks his collaborators on the cuprate project, M. Tachiki, J.-H. Chung, P. Piekarz, R. J. McQueeney, M. Yethiraj, H. A. Mook, M. Arai, S. Tajima and Y. Endoh for their contributions. He also thanks K. A. Müller, J. B. Goodenough, L. Gor'kov, A. Bussmann-Holder, D. Mihailovic, A. Bianconi, A. Bishop, P. W. Anderson, J. C. Phillips, A. Lanzara, Z.-X. Shen, W. A. Little and M. Weger for useful discussions and suggestions. This work was supported by the National Science Foundation through DMR01-02565.

References

1. e.g., Schützmann, J., Tajima, S., Miyamoto, S., Sato, Y. and Hauff, R., *Phys. Rev. B* **52** (1995) 13665.
2. e.g., McCarty, K. F., Liu, J. Z., Shelton, R. N. and Radousky, H. B., *Phys. Rev. B* **41** (1990) 8792.
3. Pintschovius L. and Reichardt, W., in *Physical Properties of High Temperature Superconductors IV*, ed. D. Ginsberg (Singapore, World Scientific, 1994) p. 295.
4. Egami, T., Chung, J.-H., McQueeney, R. J., Yethiraj, M., Mook, H. A., Frost, C., Petrov, Y., Dogan, F., Inamura, Y., Arai, M., Tajima S. and Endoh, Y., *Physica B*, **62** (2002) 316.
5. Chung, J.-H., Egami, T., McQueeney, R. J., Yethiraj, M., Arai, M., Yokoo, T., Petrov, Y., Mook, H. A., Endoh, Y., Tajima, S., Frost C. and Dogan, F., *unpublished.*

6. Toby, B.H., Egami, T., Jorgensen J.D. and Subrmanian, M.A., *Phys. Rev. Lett.*, **64** (1990) 2414.
7. Lyddane, R. H., Sachs, R. G. and Teller, E., *Phys. Rev.* **59** (1941) 673.
8. Tachiki M. and Takahashi, S., *Phys. Rev. B* **38** (1988) 218; *Phys. Rev. B* **39** (1989) 293.
9. Ginzburg, V.L., *Sov. Phys. JETP* **20** (1965) 1549.
10. e.g., von Hippel, A. R., *J. Phys. Soc. Japan* **28** (Suppl.) (1970) 1.
11. Cohen, R.E., *Nature* **358** (1992) 136.
12. Resta, R., *Rev. Mod. Phys.* **66**, (1994) 899.
13. Resta, R., *Ferroelectrics* **136** (1992) 51.
14. King-Smith, R. D. and Vanderbilt, D., *Phys. Rev. B* **47** (1993) 1651.
15. Su, W.P., Schrieffer, J.R. and Heeger, A.J., *Phys. Rev. Lett.* **42** (1979) 1698.
16. Piekarz, P. and Egami, T., *unpublished.*
17. Egami, T., Ishihara, S. and Tachiki, M., *Science* **261** (1993) 1307.
18. Resta, R. and S. Sorella, *Phys. Rev. Lett.* **74** (1995) 4738.
19. Tachiki, M., Machida, M. and Egami, T., *unpublished.*
20. Weger, M. and Birman, J.I., *Laser Physics* **12** (2002) 1.
21. Holcomb, M.J., Collman, J.P. and Little, W.A., *Phys. Rev. Lett.* **73** (1994) 2360.
22. Little, W.A., Collins, K. and Holcomb, M.J., *J. Supercond.* **12** (1999) 89.
23. McMillan, W., *Phys. Rev.* **167** (1972) 615.
24. Schrieffer, J.R., *Theory of Superconductivity* (Addison-Wesley, Reading, MA, 1968).
25. Nozieres, Ph. And Schmitt-Rink, S., *J. Low Temp. Phys.* **59** (1985) 195.
26. Tranquada, J.M., Sternlieb, B.J., Axe, J.D., Nakamura, Y. and Uchida S., *Nature* **375** (1995) 561.
27. Egami, T. and Louca, D., *Phys. Rev. B* **65** (2002) 094422.

AUTHOR INDEX